T0146251

Nested Ecology

Nested Ecology

The Place of Humans in the Ecological Hierarchy

EDWARD T. WIMBERLEY

Foreword by John F. Haught

The Johns Hopkins University Press

Baltimore

© 2009 The Johns Hopkins University Press
All rights reserved. Published 2009
Printed in the United States of America on acid-free paper
9 8 7 6 5 4 3 2 1

The Johns Hopkins University Press
2715 North Charles Street
Baltimore, Maryland 21218-4363
www.press.jhu.edu

Library of Congress Cataloging-in-Publication Data
Wimberley, Edward T.
Nested ecology : the place of humans in the ecological hierarchy / by Edward T. Wimberley ;
foreword by John F. Haught.
    p. cm.
Includes bibliographical references and index.
ISBN-13: 978-0-8018-9156-4 (hardcover : alk. paper)
ISBN-10: 0-8018-9156-6 (hardcover : alk. paper)
ISBN-13: 978-0-8018-9289-9 (pbk. : alk. paper)
ISBN-10: 0-8018-9289-9 (pbk. : alk. paper)
1. Human ecology. 2. Conservation of natural resources. I. Title.
GF41.W54 2009
304.2—dc22   2008034008

A catalog record for this book is available from the British Library.

*Special discounts are available for bulk purchases of this book. For more information, please
contact Special Sales at 410-516-6936 or specialsales@press.jhu.edu.*

The Johns Hopkins University Press uses environmentally friendly book materials, including
recycled text paper that is composed of at least 30 percent post-consumer waste, whenever
possible. All of our book papers are acid-free, and our jackets and covers are printed on paper
with recycled content.

Contents

Foreword, by John F. Haught   vii

Preface   ix

1   Developing a Practical and Sustainable Ecology   1

2   Personal Ecology   14

3   Social Ecology   32

4   Environmental Ecology   50

5   Cosmic Ecology and the Ecology of the Unknown   67

6   Essential Characteristics of Nested Ecology   119

7   The Fundamentals of Nested Ecological Householding   146

Epilogue   196

References   207

Index   241

Foreword

....

Scientists in all fields are now relying increasingly on images of the world as a complex emergent system "nesting" multiple subsystems. Nature consists of communities within communities. It is hard for any finite mind to hold all the levels together simultaneously, but in the intellectual world, and in culture at large, holistic pictures of a dynamic, evolutionary, and richly layered universe are slowly and irreversibly replacing the more mechanistic, static, vertical, and linear models that have been foundational to modern thought.

One might assume, therefore, that environmental studies would, of all fields, be the most enthusiastic about new depictions of nature in terms of the nesting model. In a sense this assumption is well founded, but rarely is an environmental thinker or ecologist sufficiently sensitive to all the dimensions that a truly integral vision of nature requires. Most environmental writers defend one level at the expense of others, and of course one can always learn much from these biased perspectives. But after reading the works of even the most skilled ecological scientists, ethicists, and philosophers, the impression usually still remains that something important has been left out. No matter how broadminded an ecologist's or environmentalist's intentions may be, readers will find that some level — whether physical, personal, social, biological, cosmic, or spiritual — has been slighted.

Terry Wimberley, however, is exceptionally aware of this almost inevitable shortcoming of ecological writing and rhetoric. His wise, readable, and convincing book awakens us to spheres of concern that even the most sensitive ecological treatises have often ignored or underemphasized. He has in mind a much more integral and nuanced ecological vision than is customary. Readers of many backgrounds and interests will find herein a carefully coordinated range of reflections on the multiple nesting and nested levels that make up the universe. Wimberley's sophisticated study of the plurality of ecological strata challenges us to develop a wider ecological awareness than even some of the most celebrated ecological visionaries have provided.

This work demonstrates persuasively, though without having to resort to homiletics, that ecological thought and ethics can be considerably enriched by taking into account the insights of psychologists, economists, politicians, sociologists, cosmologists, philosophers, and even religious thinkers. Especially satisfying, at least to me, is the author's realization that a deeply influential ecological ethic cannot take root apart from considering the importance of a "personal" ecology at one pole of the environmental spectrum and an "ecology of the unknown" at the other. For many persons only the sense of an *ultimate* environment, a dimension of endless mystery, can truly enliven and sustain their personal hope that nature still has a future.

It now seems most inappropriate, therefore, for environmental ethicists to ignore the intuition so many people have that the empirically available world is "nested" finally by an inexhaustible depth of being, value, futurity, truth, and beauty. Ethical incentive can flourish most naturally and spontaneously when persons understand that their actions and attitudes have a bearing not only on the physical, social, biological, and cosmic spheres of being but even on the ultimate environment in which they live and move and have their being. I am optimistic that readers of this fine book will find their own vision of the world and its promise expanding with each chapter.

John F. Haught, Ph.D.
Senior Fellow, Science & Religion
Woodstock Theological Center
Georgetown University

Preface

....

Imagine a set of Chinese boxes, beginning with one just large enough for you to stand in and nested within a series of progressively larger boxes stretching out into infinity. If you suppose that if you are in the middle of this set of boxes the world revolves around you and your kind, then you miss the point of this analogy. Your position in that small, central, nested domain is not a statement of your centrality — or that of humankind — to the rest of the world. It simply reflects the pragmatic reality that as a human being you can only perceive the world from your own subjective experiences, taking into account your biologically derived perceptual and cognitive capacities, and you can only interact with those progressively more expansive ecological domains through the use of your body and the tools that your body is capable of fashioning.

Your position in this nested ecological matrix is no more than your biological and perceptual niche for viewing and interacting with the world. Every other creature occupies a box of their own in this nested hierarchy, and their position no more implies that they are the masters of the universe than yours does. In fact many, many of the planet's living entities will occupy boxes much smaller than yours. However, every creature, as they perceptually and physically interact with the expansive ecological domains around them, will of necessity do so from the perspective of their own unique form of existence and place on the planet.

In presenting this book for your consideration, I would like to thank six of my intellectual mentors who have taught me to think in terms of systems theory. The first is Herbert Simon, who was writing and lecturing at nearby Carnegie-Mellon University when I was a graduate student in public affairs at the University of Pittsburgh. My lifelong research interest has been systems theory as it is applied clinically, organizationally, socially, and ecologically. I am indebted to Simon for introducing me to hierarchy and systems theory from the outset as a set of nested Chinese boxes — an image that I develop in some length in this book.

I am also indebted to Amitai Etzioni, whose work on decision making I studied during my graduate years. It was Etzioni who challenged me to think about the various ways in which decisions are made in social systems and who introduced me to his "mixed-scanning" approach. However, I am most immediately indebted to Etzioni for reminding me anew of Simon's analogy of hierarchy as a set of Chinese boxes. Without him, I might have overlooked this beautiful metaphor for living ecologically in the world.

I would also like to thank three noted family clinicians — Salvador Minuchin, Murray Bowen, and Monica McGoldrick — whose approaches to family systems theory I not only studied assiduously but also practiced for many years as a psychiatric social worker in family therapy. Minuchin, Bowen, and McGoldrick taught me not only to perceive individuals as members of current and historical family systems and family system networks, but also to conceive of family systems interacting with a complex array of social and organizational systems such that I was able to understand how behavior and values learned in families are ultimately reflected in organizational and social life. Moreover, they sensitized me to the fact that each member of a system perceives the system in a completely unique fashion, such that reality within any system is entirely a function of the person perceiving it. I have taken this insight into my work in ecology and have concluded that not only is an anthropocentric perspective on the world unavoidable, it is necessarily personal, such that every individual actor within nested ecological systems has a unique perspective on the world and their place within it.

Finally, I would like to thank Aldo Leopold for helping me understand what it means to "think like a mountain," which for me meant realizing that while every human being necessarily perceives and interacts with the world from the perspective of their own individual nested Chinese box — what I like to think of as their personal "household" — so does every other creature on the planet, and if a mountain could really think, so could ecosystems or perhaps even the biosphere itself. Leopold's insights have allowed me to apply everything else I have learned about systems theory to the level of natural ecosystems and the biosphere. I am deeply indebted to him for this insight, as I am to all of my intellectual mentors.

I trust that in perusing the pages to follow, the reader will be patient with me and willing to endure some of the more densely presented content in the interest of ultimately arriving at a fresh new ecological perspective where ecologies become households, nested and extending from the immediacy of the personal to the unfathomable and unknowable far-flung nether regions of the cosmos.

Nested Ecology

ONE

Developing a Practical and Sustainable Ecology

....

Sustainability Challenges
....

One of the truly frustrating challenges confronting those interested
in preserving and protecting the planet's fragile natural ecosystems is how to go about inculcating a sense of ecological responsibility
within individuals that will be predictably reflected in their behavior. Even
the most strident environmental activist recognizes the inherent difficulties in consistently acting in an eco-friendly fashion in today's modern
human culture. At the heart of this difficulty are the inherent contradictions to be found in functioning as an "economic" versus an "ecological"
person.

The dominant economic paradigm, based upon the philosophical
foundations of Adam Smith, Friedrich von Hayek, and Milton Friedman,
tends to conceive of nature as embodying a set of natural resources that can
be legitimately utilized to realize human wants and desires. From this
perspective, economics is perceived as a free and open democracy within
which individuals can express their wants, desires, and preferences via the
unfettered exercise of personal discretion regarding goods that will be
produced and purchased in the marketplace. This philosophy tends to
conceptualize natural resources as being either limitless in quantity or
endlessly replaceable by alternative and more plentiful resources given
ongoing developments in human knowledge and technology. Consequently, the natural world becomes a "given" — an assumed foundation
upon which seemingly endless economic activity can be predicated.

Granted, this "free-market" orientation to economics is far from the
only philosophical approach operative within the world today. However, it

1

is the dominant perspective within both developed and so-called emerging economies and is certainly the dominant paradigm influencing the daily business environment in the world's largest economies. Moreover, I would assert that this is the dominant philosophical assumption informing the typical consumption habits of the average person in the world's most affluent societies. As far as the average consumer is concerned, virtually all resources requisite for providing the goods and services demanded by today's modern society are available in seemingly endless quantities.

From an ecological perspective, it is probably an underappreciated reality that, in the interest of survival, all human beings must necessarily participate in economics and that these activities unavoidably result in the consumption of natural resources. Meeting basic daily needs such as nutrition, housing, and safety require every human being to act economically. Such economic activity requires of every person that they "produce" something of monetary value that they can sell in the marketplace — be it a good, product, or service — as well as "consume" the products of others. The creation of these consumables directly or indirectly involves transforming natural resources into products (so-called value-added natural resources) for which the market suggests there is demand given the consumptive behavior of the public.

This consuming "public" is not some hypothetical aggregate without a face or identity. The public consists of individuals who are each and every one moral "actors" within economic and ecological arenas. Even the youngest, frailest, and most infirm within the society engage in productive and consumptive activities or have someone who engages in them on their behalf. Being an economic person is not optional. It is an unavoidable function of human existence and culture. For this reason, any sustainable ecological ethic must recognize the primacy of economics in human behavior and culture and must ultimately seek to integrate ecological and economic (i.e., preference utilitarian) values at the individual level (Singer, 1993).

At the theoretical level, great strides have been made in articulating what is now referred to as "sustainable economics" (Daly, 1997; Rao, 2000). However, these conceptual models are articulated at the societal level, while consumptive behavior is an individual phenomenon. Consequently, while the development of sustainable approaches to economic behavior such as those proffered by Rao (2000) and Daly (1997) may serve to eventually transform the economic philosophies of government and industry, thereby transforming markets, the theoretical literature espousing such economies does not directly or immediately influence or inform the consumptive and productive practices of individuals.

This book is concerned with the development of ecological values at the individual level and seeks to articulate a practical and intuitive philosophy that enables people to consistently "live ecologically" in the broadest sense of the term. It does so by making a number of seemingly obvious yet previously unarticulated assumptions about human behavior and motivation. It starts with the basic assumption, originally articulated by Abraham Maslow (1968), that human beings act first to secure basic needs for safety and sustenance before seeking to realize higher-order desires for affiliation, recreation, and self-realization. Higher-order experiences, as formulated by Maslow, are built upon the foundations of having successfully realized lower-order needs and wants, and meaningful social interactions and experiences are predicated upon the successful a priori realization of self-identity and purpose.

Though many within the ecological community rail against the excessively anthropocentric nature of modern human culture, the inescapable reality is that human beings must perceive the natural world through human eyes, and by way of the prisms of human cultures. Human interactions with the natural world are inextricably mediated via human senses, human experiences, and human culture. This is the only way in which human beings can relate to the world, and it is naïve to expect people to do otherwise. As a result, any fruitful attempt to instill a meaningful, sustainable, and operational ecological ethic among individuals must begin with a complete reconceptualization of what we mean when we speak of ecologies and of the human place within them.

I would assert that in popular parlance, the term "ecology" has become virtually synonymous with "nature" or the "natural environment." Such an immediate word association has, I believe, served to oversimplify what ecology truly entails and has contributed to the perception that whatever ecology "is," humankind has somehow been excluded from it. In reality, nothing could be further from the truth. Not only are humans necessarily embedded within natural ecologies, they are also embedded within an intricate network of personal and social ecologies that are predicated upon the existence of a vast and complex web of natural ecologies. Far from being excluded from natural ecological systems or, for that matter, having excluded themselves from such systems, human beings exist only as a function of these natural ecologies. Indeed, human existence and human culture are productive expressions of more basic and fundamental natural ecosystems upon which all life depends. Human identity, society, and culture ultimately find themselves nested within a complex interdependent hierarchy of ecosystems. Thus, to write about ecology and humankind's relationship to and responsibility for natural ecologies is to speak of the relationship between a set of nested ecologies in which human identity, behavior, culture and society, are conceived of as ecological expressions and entities.

Nested Ecologies

Sustainable personal and social ecologies necessarily serve as the foundation for a sustainable environmental ecology. Such an assertion regarding the necessity for developing personal and social ecologies that are supportive of an environmental ecological ethic is one that has been underemphasized in the ecological, philosophical, ethical, and theological literature. It is my intent in this book to extend Maslow's "hierarchy of needs" (Maslow, 1971, 1968) to the ecological realm by developing a set of interrelated and "nested" ecological domains that are mutually reinforcing and supportive of one another. To that end I will introduce a concept I call "nested ecology," which refers to an interlocking set of systems that begin at the level of the self and progressively extend to encompass families, groups, communities, ecosystems, the biosphere, and beyond into the unfathomable reaches of the cosmos.

In asserting the need for a nested ecological perspective, I interpret the term "ecology" as consisting of an integrated whole progressively based upon personal, social, environmental, and cosmic ecologies. Unlike others who narrowly interpret the word "ecology" to primarily involve environmental ecology, I argue that achieving a sustainable environmental ecological ethic presupposes sequentially achieving sustainable ecologies within each of these other domains. Moreover, I will assert that ultimately all ecological domains are subsumed and embedded within a "cosmic ecology" that we are just beginning to understand, which in turn is embedded in an "ecology of the unknown" that many will relate to in terms of spirituality and religion.

From a uniquely human perspective, all ecological foundations begin with "personal ecology." Of the four primary forms of ecology dealt with in this book, personal ecology is conceptually the least developed. While there are numerous interpretations of the term available, perhaps the simplest and most compelling definition comes from Patrick Magee (1998, 242), who defines personal ecology as "the ongoing process of keeping your life in balance." Personal ecology is best described within the social and developmental psychology literature, where it is alternately referred to as "self-ecology" or the "ecology of self" (Hormuth, 1990, 1–3). As such, it involves developing a sense of "self-identity" and a "sense of place" based upon a high degree of social and environmental continuity.

Unlike personal ecology, "social ecology" is by definition family- and community-oriented. Indeed, it is as sociological and economic in nature as personal ecology is psychological. While numerous interpretations of "social ecology" will be discussed in the book, all are grounded in the thought of Murray Bookchin (2001, 1996, 1994, 1992). Unfortunately, Bookchin's perspective regarding social ecology is primarily characterized by what social ecology *is not* rather than by what it is (Zimmerman, 1997). Bookchin's philosophy reflects his distinctly anarchistic, communistic, and antiestablishment political orientation, thereby rendering it unsuitable as a theoretical foundation for creating a functional approach to social ecology. Thankfully, more contemporary communitarian approaches to social ecology have been developed by such thinkers as John Clark (1990), Amitai Etzioni (1998), J. Baird Callicott (1996), and Wendell Berry (1996) that are more useful in terms of integrating personal and environmental ecological needs within a dynamic social milieu.

By comparison, environmental ecology (the ecological concept most frequently considered as synonymous with the term "ecology") refers to "nature" and natural environments and ecosystems. Unfortunately, environmental ecology is too frequently expressed as an "ecology of not," or in terms of what threatens the environment and what needs to be done to remediate environmental harm rather than articulating an "affirmative ecology" of what an environmental ecology should entail. Fortunately Donald Scherer (1990), with his "upstream" and "downstream" ecological model, provides a perspective that lends itself to an affirmative and sustainable vision of environmental ecology, as do the so-called deep ecologists Arne Naess (1989, 1973), Bill Devall and George Sessions (2001), and Holmes Rolston III (1988).

Ultimately, all nested ecological realms are subsumed within a cosmic ecology of which the Earth's biosphere is the basic unit. The biosphere is the home for all known life and is the entity through which all creatures and ecosystems interact with the rest of the universe. As Vladimir Vernadsky has so presciently observed in his important book *The Biosphere* (1926), all life embodies energy emanating from the sun and from throughout the cosmos. Consequently, the Earth as biosphere is the repository and reflection of the influence of the cosmos upon its environs, ecosystems, and life forms. Practically speaking, that means that all nested ecosystems culminate in and are dependent upon a cosmic ecology upon which all life on the planet depends.

Ecology as an Evolving Concept

While I refer to ecology in terms of a set of nested and interdependent hierarchies, it is necessary also to understand ecology as an evolving concept. Ernst Haeckel first used the term in *General Morphology of Organisms* (1866) where he defined ecology as "the relation of the animal both to its organic as well as to its inorganic environment." In coining this term, Haeckel sought to describe the study of interrelations among organisms that influence organism form and function. Given the preceding discussion of nested ecologies, it is obvious that the term has been greatly expanded owing to the influence of systems theory (discussed below) of which Haeckel was an early pioneer. Today the terms "ecology" and "environmental ecology" are used interchangeably, with ecology alternately defined as "the scientific study of the processes influencing the distribution and abundance of organisms, the interactions among organisms, and the interactions between organisms and the transformation and flux of energy and matter" (IES, 2005), or as "the relationship of living things to one another and their environment, or the study of such relationships" (U.S. EPA, 2006).

Ecological inquiry has been significantly enhanced by the growth in influence of "general system theory," which originated in the work of Kenneth Boulding (1956) and Ludwig von Bertalanffy (1968). Boulding's approach to general systems theory was in turn expanded to include economic and social systems theory (Parsons, 1971; Stigler, 1946). More recently, Littlejohn (1983) applied systems theory to environmental studies, having defined a system as "a set of objects with attributes that interrelate in an environment," and characterized these attributes as including the "qualities of wholeness, interdependence, hierarchy, self-regulation, environmental interchange, equilibrium, adaptability, and equifinality" (Littlejohn, 1983, 32). Implicit within the theory is the view that systems are imbedded within one another in a hierarchical fashion, such that changes in one portion of the system can be felt throughout other systems. Littlejohn's perspective is reflected in the model of nested ecology presented in this book.

The hierarchical nature of systems implies that higher-order systems consist of and are dependent upon the functioning of lower-order systems. Consequently, complex system functions are dependent upon the work of comparatively less complex systems, with all systems intertwined in one self-regulating totality. This perspective conflicts with social ecology's traditionally anti-hierarchical orientation, which narrowly associates hierarchy with "top-down" systems of authority and power. A systems theory orientation toward ecology is particularly important to the thesis of this book, since it both supports the concept of a "hierarchy" of system needs and provides a theoretical rationale for the nested hierarchical organization of ecological systems (personal, social, environmental, and cosmic) presented herein.

"Nested hierarchy" is an emerging concept in psychology, biology, and ecology. This concept has been most prominently applied in the fields of psychology and human ecology by Urie Bronfenbrenner (1978), whose work is grounded in the theoretical research of psychologist Kurt L. Lewin (1931). Lewin and Bronfenbrenner conceive of human ecological environments or systems as consisting of sets of "nested and interconnected structures," beginning with *microsystems,* which includes very basic human groupings such as the family or a classroom, *mesosystems,* to include any set of interacting microsystems (or a system of microsystems), *exosystems,* which involve settings in which a particular "evolving person" is not present, but that influence the evolving person, and *macrosystems,* which includes the larger cultural context surrounding all systems (Bronfenbrenner, 1978, 203–5).

While I find Bronfenbrenner's concept of nested hierarchy (Bronfenbrenner, 2005) to be extremely useful as a human-oriented, psychological, and social systems theory, I find it excessively anthropocentric in its orientation to be completely useful within the context of this book. Consequently, I favor a more biologically oriented nested hierarchy model approach — one that operates on the principle that larger and more complex systems consist of and are dependent upon simpler systems and essential system-component entities. Such an approach has not only been applied in a biological and ecological sense (Hewlett, 2003) but has additionally been expressed as a philosophical construct (Clayton, 2004, 87). In this fashion "atoms are in molecules, molecules in cells, cells in organs, organs in organisms" right up the system hierarchy until animals reside in ecosystems, smaller ecosystems reside within larger ecosystems, macro-ecosystems reside within planets, planets reside within solar systems and so on (Clayton, 2004, 88).

Some Implications of Construing Ecology as a Set of Nested Domains

The implications of approaching ecology as a set of nested domains are important for developing a sustainable approach to ecological values and action. For instance, in environmental education, rather than exclusively and narrowly focusing upon the natural environment as the focal point for ecological ethics and human moral action, a nested approach to ecology would suggest that discussions of ecology begin with a review of personal ecology. To that end, inquiry might start with a review of lifestyle in which individuals are asked to consider basic needs such as housing, health, nutrition, psychosocial well-being, occupation, and social support systems. It would additionally explore personal habits, values, attitudes, and behaviors in terms of the degree to which they contribute to the overall health and well-being of the individual, or alternatively serve as a detriment to such health and well-being.

Topics that might be explored in the interest of promoting personal ecology include safety, sustainable and nurturing social relationships, job/career satisfaction, spirituality, and meaningful leisure/recreation. Likewise, personal ecology would promote living harmoniously within both human and natural communities and would explore concrete ways in which such harmony could be facilitated and promoted. Attitudes and skills that might be pursued to promote social harmony include tolerance, patience, courtesy, generosity, and cooperation. Discussions of living harmoniously in the natural community could address such topics as consumption versus need, managing personal waste, transportation, and energy use.

Any attempt to promote or facilitate personal ecology should recognize that the realization of a sustainable personal ecology is largely dependent upon the individual's social relationships and networks. Those pursuing meaningful and sustainable personal ecologies must come to the realization that their very identities are grounded in the social networks from which they have emerged and in which they interact and participate. No personal ecology can be satisfying or complete unless it is thoroughly grounded in a healthy, functional, and sustainable social ecology. Consequently, promoting ecological well-being requires the exploration of social relationships and networks. This should necessarily begin with an exploration of family and family history. It should inquire as to the personal identity of the individual within the context of their family — asking the pertinent questions: "Who are you?" and "What is expected of you?" Asking these basic questions sets the stage for asking the more pertinent questions of "Who do you aspire to be?" and "How well are you living up to the expectations of others *and* of yourself?"

As is the case with personal ecology, a review of a person's social ecology should focus upon that system's health and sustainability. Efforts in this regard should ideally enable the individual to consider the extent to which their relationships with friends, family, coworkers, and so on, are sustainable, functional, fulfilling, and contributive to the ongoing health and well-being of themselves and others. Social problems and impediments should be identified, and the individual should be invited to consider the nature of their personal contributions to these difficulties, as well as the contributions of others. Such a review should be based upon the recognition that some social systems and relationships are inherently "healthy" — that is, they promote the physical and emotional health and well-being of all those involved in them — while others are inherently "unhealthy" and even destructive. Consequently, one of the central goals involved in reviewing any social ecology should be to identify areas in which social systems and relationships are healthy and ways to sustain these systems and relationships, while simultaneously identifying system and relationship dysfunctions and engaging the individual in the process of remedying these problems.

Although this is a necessary step, it constitutes only an initial step in the process of exploring one's social ecology. Beyond the immediate web of social relationships that individuals involve themselves in, exploration should additionally extend to the community level, inquiring as to the quality of life to be found in a person's neighborhood, town, city, or area. Such an inquiry should seek to identify the manner in which the community fulfills the needs of its members as well as discerning areas of community dysfunction. In exploring this aspect of social ecology, the individual should be encouraged to consider the nature of their "civic" responsibilities and to determine how they might go about addressing these responsibilities and contributing to the health and welfare of their community.

Moreover, the individual should be encouraged to construe their community's well-being as contingent upon the welfare of other communities — both near and far. In this regard, economic, political, cultural, and environmental considerations should come into play. Just as no individual has a personal identity independent of the social milieu, likewise no community exists in isolation from other communities. Therefore, it becomes incumbent upon the individual not only to appreciate the need to sustain and promote the health of "communities of communities," but to act in a civic fashion to do so. Again, the focus of ecological inquiry is "health and well-being," not simply on a temporary basis but in the interest of sustaining the community as a healthy entity over the long run. Civic attitudes and actions devoted toward the immediate or short-term welfare of the community, or actions grounded essentially in self-interest are not to be discouraged or denigrated. However, the individual should additionally be challenged to consider the welfare of the community beyond their own lifetimes, and to commit themselves to actions and attitudes that promote sustainable communities.

Unavoidably, a portion of the effort required to sustain and promote communities must address the relationship between human communities and the larger natural environs in which these are embedded. In this fashion, a consideration of personal and social ecologies leads to a consideration of natural environments and environmental ecology. Fortunately, environmental educators, policy advocates, and activists are well versed in how to mentor individuals into assuming a more environmentally sensitive and responsive lifestyle, so I will not describe this process in any detail. I would, however, argue that the major impediment to successfully motivating people to sustain an environmentally appropriate lifestyle involves their penchant to prioritize anthropocentric preferences and desires over ecological responsibilities. This is the principle reason why I developed a nested approach to ecology, because I recognized that for an environmental ecology to be both meaningful and sustainable, it simply must be grounded in personal lifestyles and in human communities that are also meaningful, healthy, and sustainable.

Such an approach arguably expands the study of ecology beyond a narrow focus upon the environment to considerations of human life and society within the context of the natural environment. In fact, it could be asserted that such an approach makes a study of ecology virtually a study of life and living itself. A nested ecology is devoted not simply to the study of "natural" life, "species" life, "human" life, or the life of "ecosystems," it entails considering all life and all that is required to maintain and sustain it. While this nested approach may offend some because of its decidedly anthropocentric orientation, this approach is necessary and purposeful in the interest of engaging people in the process of reframing existence as something that necessarily occurs within the context of a complex system of animate and inanimate ecosystems.

Environmental ecology, moreover, is not the ultimate dimension of ecology that needs to be considered. The planet Earth is but one of billions of planets and stars that we know to exist throughout the universe. This is a realm that we know little about, but what we know is more than sufficient to suggest that just as Earth embraces a seemingly endless number of ecosystems, so does the universe embrace an endless number of cosmic ecosystems. The implications of this relationship are only beginning to be understood, but it stands to reason that if the health of the planet is dependent upon the health and well-being of its constituent ecosystems, then the health of the cosmos must also be dependent upon the integrity of its constituent ecosystems. Currently this assertion cannot be definitively validated, but it would seem that what is known about terrestrial ecosystems must be to some extent applicable to cosmic systems.

Admittedly, while humans seem to have the capacity to utterly destroy life for themselves on Earth, it does not seem that we are anywhere near capable of causing any significant damage to another planet — let alone being of any significant threat to the solar system, the galaxy, or the universe. Those who envision such a potential delve into the world of science fiction and do not reflect any current or anticipated reality. However, just because we don't appear to be an eminent threat to the cosmos does not mean that we can take cosmic ecosystems for granted. Increasingly, the sense among people of science and spirituality is that the universe is an integral component of the biospheric ecosystems that we inhabit and depend upon. Consequently, as a matter of principle, these cosmic ecological systems are entitled to the same deference owed terrestrial ecosystems in the interest of vouchsafing their ongoing integrity and sustainability.

Such a realization may lead some to wonder if there exists "universes of universes" just as there are "communities of communities." Likewise, they may wonder how all that is came into being and whether or not there is a force or presence behind everything creative and influential in the affairs of the planet and the cosmos. Fueled by the unknown, yet inspired by what is known or virtually known, these individuals will see yet one final realm of ecology beyond the primary set of three ecologies encompassed by the limits of the planet. For them, the most encompassing ecological realm — encompassing the entire universe *and* the lives and communities of humans, living species, and all of their environmental environs — is that of spiritual ecology. Spiritual ecology is based upon conjecture, doubt, faith, and belief. It is not grounded in what is known but is rather a response to the unknown and the desire to know. Those embracing such a vision of the universe may do so within the confines of religious faith and belief or may do so in a more loosely defined "spiritual" fashion. They may construe this realm as the product of a specific divinity or divinities or may perceive a sense of mysticism and spirituality within the nature of the cosmos itself.

Those who share such an ecological perspective may be criticized as being insane, deluded, "in denial," or as coping naïvely with the finitude of human existence. Adherents of a spiritual ecology may feel called upon or constrained by the divine to adhere to a discipline and lifestyle regarding the other nested ecology realms and may construe their entire existence as being dedicated to these normative expectations. Alternately, their sense of spirituality might be expressed and nurtured in noninstitutional settings, and their expectations of themselves and others may be poorly defined or undefined altogether. Regardless of how spiritual ecology is construed or realized, there is no doubt that this can be a powerful paradigm that can provide a sense of identity, place, and meaning in the lives of people and serve as a powerful motivator toward ecologically ethical thought, values, and action. While it is not a "necessary" ecological perspective for all people, for those who perceive the world through spiritual or religious eyes, it is not only necessary but mandated by their beliefs.

These examples suggest that approaching ecology from a nested perspective results in a very different and much more comprehensive approach to the study and practice of ecology. Ecology is much more than "environment" or the study of the environment. As will be seen in the chapters to follow, ecology is about "home," "place," "households," and what I like to call the process of "householding." Ecology is not just a way of life, it is life itself and pertains to every fashion in which humans live as well as to a wide array of other "households" and communities. Approaching ecology on any lesser scale is to insure that the battle for ecological sustainability is lost, and with it, humanity's place in the world. I encourage the skeptical reader to approach the pages to follow with an open mind and to allow for the possibility that ecological sensibility and action are best achieved by considering all human life and experience as an ecological event.

TWO

Personal Ecology

....

Defining Ecology
....

Ernst Haeckel (1866) coined the term "ecology" to describe the study of interactions or relationships among organisms that influence their organismic form and function. This popular term is of great antiquity. Ecology is a derivation of the Greek word, *oikos,* which literally means "home," "household," or "a place to live" and was the principal concern of the ancient Greek philosopher Theophrastus (Anton, 1999). So conceived, to be an ecologist is to be a "householder," and acting in an ecological fashion is the equivalent of what I call "householding." "Householding" involves making a home in the world and taking care of all those duties and responsibilities that are associated with establishing and maintaining a household and home. Narrowly conceived, householding (personal ecology) entails taking care of oneself. More broadly conceived, it involves caring for one's community (social ecology), and at even broader levels, householding involves caring for the planet (environmental ecology), caring for the universe (at least our corner of the universe, as in the case of cosmic ecology), and honoring the divine (spiritual ecology).

Personal Ecology versus Human Ecology
....

This chapter concerns the primary realm for ecological identity: personal ecology. Personal ecology is conceptually similar to but distinctly different from the related discipline of human ecology. Central to the concept of human ecology is the understanding that humans are also animals. Consequently, human ecology focuses upon humans as fellow creatures sharing the planet with other creatures and entities and seeks to understand how human behavior and culture influence and transform the world's environment. Human ecology of necessity embraces a variety of disciplines, including human development, environmental ecology, and evolutionary biology (Levine, 1975).

The literature of human ecology has assumed two distinct formats. In the first, human culture is placed within the context of nature for the purpose of assessing the impact of human society upon the world. The work of such writers as Frederick Steiner (2002), David Orr (2002), Gerald Marten (2001), and Charles Southwick (1996) is illustrative of this approach. By way of comparison, the work of Urie Bronfenbrenner (2005) and his contemporaries Stephen Ceci (1996) and Robert Cairns, Lars Bergman, and Jerome Kagan (1998) emerges from the perspective of developmental psychology and human development, construing human growth, behavior, and culture from within the context of a set of nested ecological systems. While both approaches contribute to an understanding of human interactions with other ecosystems, I contend that Bronfenbrenner's human development approach is more nearly reflective of the perspective that I refer to as "personal ecology," whereas the "human ecology" perspective espoused by Orr (2002), Marten (2001), and others is more akin to the concept of "social ecology" that will be discussed in the next chapter.

In comparison to human or social ecology, personal ecology is more psychologically or psychosocially oriented. Unlike human ecology, which assumes the perspective of humanity's relationship to the environment, personal ecology is by definition "personal," assuming an individual perspective on the environmental, social, and cultural systems within which the lives of people are embedded. Personal ecology becomes important conceptually because it provides insight into what motivates, fulfills, and satisfies the needs of individuals. An underlying assumption of this book is that unless individuals find ways of creating and sustaining meaningful lifestyles (which we will discuss in terms of functioning within fulfilling personal ecological systems) there is little chance that they will contribute to sustainable social ecologies or, for that matter, even value sustainable environmental ecologies. Unfortunately, personal ecology has received less attention in the ethical and ecological literature than have human, social, and environmental ecologies. This comparative inattention demands that this chapter begin with an attempt to understand what "personal ecology" implies and how it relates to ecological ethics.

Perspectives on Personal Ecology

The term "ecology" is etymologically rooted in the Greek words *oikos* and *logos,* which, taken together, refer to household knowledge, or household pattern (Cobb, 2002). So construed, "ecology" begins in the home and involves structuring the household. Consequently, the term "ecology" necessarily entails defining one's place in the world through the process of making and sustaining a home in some particular place, locale, or community where one is rooted and through which one is sustained. Alternately, I like to think of personal ecology as involving the process of personal "householding."

While Rutledge contributes to the understanding of personal ecology by reminding the reader that ecology necessarily begins with personal ecology, the contemporary interpretation of this term is in part reflected in the management literature. Illustrative of management's orientation is a description of personal ecology provided by Patrick Magee, who asserts:

> Each life is maintained by a complex series of relationships. When something happens in one area that throws the overall system out of balance, a series of forces are mobilized to correct the situation. I use the term personal ecology to describe the ongoing process of keeping your life in balance. When engaging in an activity that threatens this balance, I say that it is not "personally ecological." Personal ecology is governed mostly by subconscious mental processes designed to help keep you alive. If you do something that is not personally ecological, these subconscious mental processes usually figure out some way to distract you from doing that activity. (Magee, 1998, 243)

For Magee, personal ecology is primarily a discipline for maintaining life equilibrium in the interest of maximizing personal and professional productivity. Moreover, personal ecology serves to maintain a balance and synergy between life purpose or purposes, personal energy, and one's capacity to effectively and efficiently apply new information to personal initiatives (Magee, 1998).

The major alternative perspective regarding personal ecology comes from social psychology, which is heavily indebted to Julian Rotter's "social learning theory." Social learning theory proposes that personality is a reflection of the interaction between an individual and their environment (Rotter, 1954). This theory was later adapted to develop a theory of "self ecology," or "the ecology of the self," which consists of three primary components (Hormuth, 1990, 1–3): (a) "other persons" — reflecting verbal and nonverbal behavior to create a self that is "the internal organization of external roles of conduct," (b) "objects" — serving as social tools for conducting social roles and contributing to the establishment of self-identity, and (c) "environments" — settings that provide a place and context for individual experiences and actions. Hormuth observes that while each of these elements has been widely discussed in the psychological literature, they have rarely been considered as a unit. When they are taken together, Hormuth believes that a "picture of an ecological system emerges" in which "self-concept exists in interdependence with its ecology of others, objects, and environments" (Hormuth, 1990, 3). He hypothesizes that as long as the ecology of the self remains stable (implying that the three basic constituent parts also remain stable) an individual's self-concept will remain intact. However, to the extent that self-ecology is disrupted, so too is a person's self-concept. Hormuth's ecological model provides a theoretical rationale in support of ecological philosopher Wendell Berry's assertion of the importance of a sense of place for maintaining self-identity. According to Berry, "Not knowing where you are, you can lose your soul or your soil, your life or your way home" (Berry, 1984, 22). Berry's perspective will be discussed in greater depth presently.

Place-Identity

Social psychology also addresses the importance of place as a component of self-identity. The concept was first developed and presented by Harold M. Proshansky (Proshansky, 1978, 1976; Proshansky, Fabian, and Kaminoff, 1983) who defined "place-identity" as a substructure of the self-identity of a person consisting of, broadly conceived, cognitions about the physical world in which the individual lives. These cognitions represent memories, ideas, feelings, attitudes, values, preferences, meanings and conceptions of behaviour and experience that relate to the variety and complexity of physical settings that define the day-to-day existence of every human being. At the core of such physical environment-related cognitions is the "environmental past" of the person; a past consisting of places, spaces and their properties which have served instrumentally in the satisfaction of the person's biological, psychological, social and cultural needs. (Proshansky, Fabian, and Kaminoff, 1983, 59)

Based upon this theory, the significance and meaning of places is mediated through an individual's social experiences, to include what others think or communicate about a given locale.

Central to the idea of place-identity is the degree to which a particular environment is conducive to stability and/or change. Individual changes are mediated by place-related cognitions, including memories directly related to changing or modifying a particular place, recollections related to physical settings in which interpersonal transactions occur, and cognitions associated with physical environments where the behaviors of others cannot be changed. These cognitions in turn relate to and demand the development of a set of necessary personal ecological competency skills:

- Environmental understanding: the ability to interpret the meaning of changes in the environment as these changes pertain to the self

- Environmental competence: the ability to act and react appropriately within the context of a particular setting

- Environmental control: the capacity to successfully modify or change the person — environmental relationship

If an individual is successful in developing these competency skills, then they can be expected to maintain a sense of psychological and social equilibrium. Typically, purely homeostatic environments do not exist in natural or social systems. Consequently, the individual must acquire the skills to cope with dynamic and ongoing change — to include periods of profound and dramatic environmental disruption.

Environmental Stability and Locus of Control

Craig, Brown, and Baum (2000) discuss how unstable home environments contribute to childhood and adult anxiety. Julian Rotter also addressed this problem in much of his later work as he developed the concept of "locus of control" (Rotter, 1989, 1966). According to this theory, individuals with a perceived internal locus of control believe that they can influence and determine behavioral rewards and reinforcements through their own actions, whereas those perceiving an external locus of control believe that they are at the mercy of forces beyond their influence such as luck or fate. Implicit in this concept is the belief that individuals have the capacity to manipulate and control their environment to meet their needs. This capacity, discussed above as an "environmental control" competency skill, is essential for individuals to experience a measure of mastery over life events.

Locus of control is, at least in part, related to a stable sense of place or role, including both social and geographical components of place-identity. For instance, a warm, encouraging, and nurturing home environment in which the child is allowed more opportunities for decision making and self-mastery appears to be more conducive to the development of a sense of internal locus of control. By comparison, persons from settings where they were able to exercise only limited autonomy or were unable to marshal significant material resources appear more likely to develop a sense of external locus of control (Rotter, 1975, 1971). The relevant research literature is unequivocal in reiterating that locus of control born out of social and environmental stability is essential for maintaining optimal health and well-being. Comparatively, weak locus of control born of social instability can negatively influence health (Bobak et al., 1998).

Necessary Characteristics of a Personal Ecology

When the social psychology literature relating to the ecology of the self is considered in its totality, and when significant emphasis is given to the importance of locus of control and place-identity in the psychosocial formation of self-identity, personal ecology (or "self ecology") can be construed as necessarily including the following characteristics:

- an individual, *personal orientation* toward nature and society;

- a *secure sense of place* or place-identity, that is geographically, socially, culturally, and spiritually grounded; and

- a high degree of *social and environmental stability*, internal locus of control, and high expectations regarding individual capacity for controlling the outcome for most routine life events.

Assuming these are the foundational needs for an individual's personal ecology, anything that serves to undercut or erode these basic needs can be construed as a threat to the development and maintenance of a sense of self within the context of the larger environment. Therefore, developing and maintaining a functional personal ecology entails coping with the myriad threats that can be expected to emerge from within and beyond the personal aegis of the individual.

Threats to Personal Ecology: Anomie and Sense of Place

Philip Slater, in his landmark book *The Pursuit of Loneliness*, asserted that American culture "struggles more and more violently to maintain itself, [but] is less and less able to hide its fundamental antipathy towards human life and human satisfaction" (Slater, 1990, 122). In this book and in a subsequent text, *A Dream Deferred: America's Discontent and the Search for a New Democratic Ideal* (1992), Slater depicts an American culture in which many are disenfranchised and exhibit a sense of isolation and meaninglessness that Slater refers to as "anomie."

This concept, originally introduced by Emile Durkheim in *The Division of Labour in Society* (1947), has now become familiar in the lexicon of post-modernism. Anomie describes a society in flux where the rules of social interaction are changing and in which people do not know what to expect of one another or themselves. Simply put, anomie refers to a state of comparative societal normlessness, in which social rules and expectations are either confused or nonexistent. According to Durkheim, this condition contributes to lawlessness and deviant behavior, as well as an increase in the rate of suicide (Durkheim, 1951). Accordingly, Durkheim hypothesized (1947) that societies tend to evolve from simple, nonspecialized (so-called mechanical) states into highly complex and specialized forms, which he referred to as "organic" states. In the process, social bonds that unite people are disrupted and what follows is the evolution of a comparatively impersonal culture.

The major threat of anomie is that it serves to "disorient" members of society regarding their appropriate role and place. Indeed, anomie erodes what ecologists refer to as "sense of place" (Stegner, 1992; Allen and Schlereth, 1991) and what social psychologists refer to as "place-identity" (Dixon and Durrheim, 2000). Quite literally, anomie refers to a condition in which members of society find it difficult to locate or orient themselves within the society. As a result, they tend to lose their sense of place and become socially, morally, and spiritually disoriented.

The Temporary Society

By the late sixties and early seventies, social theorists were beginning to actively grapple with an increasingly impersonal and mobile society where it was becoming increasingly difficult to acquire and maintain a sense of place. While some lamented the social costs associated with the new shape of modern society, others offered suggestions for creating what could later be referred to as a *portable sense of place* — a self-identity independent of where one resided. In the late sixties Warren Bennis teamed with Philip Slater to write *The Temporary Society* (Bennis and Slater, 1968), which discusses the future of democracy, the democratization of the family unit, and new paradigms for organizational development beyond traditional bureaucratic models. The text also discusses so-called temporary systems and culminates with an essay on "The Temporary Society."

Bennis characterizes temporary systems as entailing "nonpermanent relationships, turbulence, uprootedness, unconnectedness, mobility, and above all, unexampled social change" (Bennis and Slater, 1968, 124), by which he refers to patterns of change that fall outside of historical experience. In discussing what he perceives to be characteristics of a "temporary society," Bennis asserts that although it is too late to change the pace of American (and Western) culture, it is not too late to explore ways for people and institutions to adapt to current social and economic forces. To that end, Bennis recommends the following changes in the U.S. approach to public education. He asserts that the educational system ought to (Bennis, 1966, 55):

- help us to identify with the adaptive process without fear of losing our identity,
- increase our tolerance of ambiguity without fear of losing intellectual mastery,
- increase our ability to collaborate without fear of losing our individuality, and
- develop a willingness to participate in social evolution while recognizing implacable forces.

Bennis further suggests that student learning should include the following:

> Learning how to develop intense and deep human relationships quickly — and learn how to "let go." In other words, learning how to get love, to love and to lose love; Learning how to enter groups and leave them; Learning what roles are satisfying and how to attain them; Learning how to widen the repertory of feelings and roles available; Learning how to cope more readily with ambiguity; Learning how to develop a strategic comprehensibility of a new "culture" or system and what distinguishes it from other cultures; and finally, Learning how to develop a sense of one's uniqueness. (Bennis and Slater, 1968, 127)

In making these recommendations, Bennis reflects a growing sense that social forces are largely beyond human control, necessitating human adaptation to the current culture. This adaptation, though reflective of the need to develop personal ecological competency skills such as those described earlier, nevertheless requires a separation of person from place, compelling modern citizens to continually bring a portable sense of place wherever they go. It also applies to making and ending relationships and commitments, as well as rendering all attachments conditional.

Personal Ecology and Sense of Place

For some, the "temporary society" represents the worst possible characteristics of modern culture. Such a philosophy makes all relationships contingent and renders all environments impermanent. Society's "temporary" quality, it is believed, serves to reduce the strength of social and environmental commitments, further exacerbating the sense of anomie and isolation.

One influential ecological philosopher has challenged the assumptions of "The Temporary Society" by suggesting that *where* one lives is very much important. According to Wendell Berry, in his essay "Poetry and Place," "How you act should be determined, and the consequences of your acts are determined, by where you are. To know where you are (and whether or not that is where you should be) is at least as important as to know what you are doing, because in the moral (the ecological) sense you cannot know what until you have learned where. Not knowing where you are, you can lose your soul or your soil, your life or your way home" (Berry, 1984, 22). Berry's insight is illustrative of what personal ecology entails. It is largely about knowing where you are geographically, socially, historically, culturally, psychologically, and spiritually. Moreover, by discovering where they are situated, the individual is empowered to discover "who" they are and their purpose for living.

Personal ecology necessarily involves defining and redefining self within a myriad of contexts, but always begins with an understanding of a "sense of place." Unfortunately, today's culture of anomie makes it increasingly difficult for many people to acquire a sense of their place in the world. Even so, Berry insists that it is important for every person to have a sense of belonging to some particular place. While he recognizes that through the power of markets and technology the world has become globally connected, Berry eschews the possibility of any person really becoming a citizen of the world. Accordingly, he asserts,

> one cannot live in the world; that is, one cannot become, in the easy
> generalizing sense with which the phrase is commonly used, a "world
> citizen." There can be no such thing as a "global village." No matter
> how much one may love the world as a whole, one can live full in it only
> by living responsibly in some small part of it. Where we live and whom
> we live there with define the terms of our relationship to the world and
> to humanity. We thus come again to the paradox that one can become
> whole only by the responsible acceptance of one's partiality. (Berry,
> 1996, 123)

With this assertion, Berry states the obvious, that humankind is inevitably place-bound, compelled from necessity to define relationships with one another and with the Earth by virtue of our personal and societal niche on the planet.

Wallace Stegner provides an even clearer understanding of humankind's sense of place asserting that "a place is not a place until people have been born in it, have grown up in it, lived in it, known it, died in it — have both experienced and shaped it, as individuals, families, neighborhoods, and communities, over more than one generation" (Stegner, 2000, 22). Stegner's perspective reflects that of Wendell Berry and others who foresee the need for society to become less transient and more deeply rooted in productive and sustainable communities. Stegner and Berry provide a vision of sense of place that they believe is conducive of ecological sustainability. However, in doing so, both authors place humankind squarely in the midst of the world of nature that they also seek to sustain. For both authors, environmental ecological sustainability necessitates sustaining humankind, human culture, and human society. The process of sustaining humankind inevitably involves countering the forces of anomie that increasingly have defined Western society.

For Berry, countering the forces of anomie involves recreating the local community, which, he asserts, has suffered from the forces of modernization and market economies. Berry notes that

> As local community decays along with local economy, a vast amnesia settles over the countryside. As the exposed and disregarded soil departs with the rains, so local knowledge and local memory move away to the cities or are forgotten under the influence of homogenized sales-talk, entertainment, and education. This loss of local knowledge and local memory — that is, of local culture — has been ignored, or written off as one of the cheaper "prices of progress," or made the business of folklorists. Nevertheless, local culture has a value, and part of its value is economic. (Berry, 1990, 157)

Berry and Stegner believe that protecting and promoting local communities is a prerequisite for creating a sense of place that not only sustains people, but also fulfills the widest possible range of human needs. The only caveat to their vision is that *human wants* must be differentiated from *human needs,* and that needs and wants must be sustainably provided given the limited resources to be found within and around local communities.

As stated at the outset of this chapter, one of the major themes of this book is that individuals and communities can only achieve a sustainable environmental ecology by first attending to needs found within personal and social ecologies. Such needs are understood as being hierarchical in nature (as described by Hewlett, 2003; Maslow, 1971), with the fulfillment of basic needs serving as foundations upon which the satisfaction of higher needs depends. Chief among these basic needs, perhaps at the level of basic safety, is the need for people to be able to "locate" themselves within the society and achieve a sense of place. As Stegner's definition of sense of place implies, locating oneself in the society is not as simple as acquiring a geographical orientation. Sense of place also involves locating oneself within a variety of other contexts. This text asserts that the central task of acquiring a sustainable personal ecology (i.e., an ecology that fulfills and sustains an individual throughout a lifetime) is the task of "self-location," or acquiring a sense of place in the world.

Modern Nomads: Creating a Portable Personal Ecology

It would appear that the pragmatic solution to coping with the modern world provided by Bennis and Slater in *The Temporary Society* is completely at odds with the community and place-centered approach of Wendell Berry. On the one hand, Bennis and Slater encourage citizens to adapt to a changing world (or become overwhelmed by it). By comparison, Berry's philosophy is much more radical and dares society to question the basic assumptions behind our modern world. From the perspective of the individual person who lives most immediately and intimately in the world within the confines of her or his personal ecological space, there is a sense of being caught up in something too large to understand or influence. For many, adaptation is the only conceivable option. Even so, the face-validity of Berry's call for local community is alluring. At issue is whether his vision is actually achievable given the demands of society upon individuals to maintain a portable sense of place and self. Likewise, one can only wonder whether human beings can functionally maintain such portability on an ongoing basis — forever abandoning any expectation of permanence.

The bitter reality is, however, that individuals existing within modern Western-styled societies truly exist within a social, psychological, cultural, and economic milieu that is increasingly temporary and rootless — a permanent state of flux and impermanence. Consequently, in these modern times, it is ethically legitimate to question whether anyone should ever accept the "temporary" and transient nature of modern society as a given and not challenge its basic assumptions. While it is pragmatic and to some extent even necessary to adapt to the "temporary society," it is questionably ethical to concede to the powers-that-be a world in which deriving a sense of place (in the concrete, community-bound sense of the term) is deemed to be an unreasonable expectation. Moreover, given the research presented from social psychology, it is clearly unhealthy and dysfunctional not to pursue the creation and restoration of community.

Communitarians, the Land Ethic, and Sense of Place

In recent years, communitarianism has emerged as an important movement seeking to address many of the deficiencies associated with the temporary society. Consider, for instance, the preamble of the Responsive Communitarian Platform:

> American men, women, and children are members of many communities — families; neighborhoods; innumerable social, religious, ethnic, work place, and professional associations; and the body politic itself. Neither human existence nor individual liberty can be sustained for long outside the interdependent and overlapping communities to which all of us belong. Nor can any community long survive unless its members dedicate some of their attention, energy, and resources to shared projects. The exclusive pursuit of private interest erodes the network of social environments on which we all depend, and is destructive to our shared experiment in democratic self-government. For these reasons, we hold that the rights of individuals cannot long be preserved without a communitarian perspective. (Communitarian Network, 1991, 1)

For communitarians, the "place" in sense of place is predominantly the local community. In this regard the movement has much in common with the ideas of Wendell Berry. Communitarians are particularly concerned about the support and revitalization of informal social institutions (often voluntary, nonprofit, and religious) that mediate community values and provide a nongovernmental network of services, activities, and support that help define the fabric of community identity (Etzioni, 1998). However, the communitarian sense of place is not entirely defined within the bounds of human community. It is increasingly defined in terms of natural biotic communities. Chief among those associated with this environmental communitarianism is the disciple of Aldo Leopold, J. Baird Callicott (1996).

Callicott champions Leopold's land ethic, which states that "a thing is right when it tends to preserve the integrity, stability, and beauty of the biotic community. It is wrong when it tends otherwise" (Leopold, 1949, 224–25) and observes that the most significant feature of the land ethic is "that the good of the biotic community is the ultimate measure of the moral value, the rightness or wrongness, of actions" (Callicott, 1980, 318). Within this perspective, personal ecology grounded in a community-based sense of place is morally trumped by the welfare of the greater biotic community (of which humans are considered a member). Callicott further asserts that "In every case the effect upon ecological systems is the decisive factor in the determination of the ethical quality of actions" (Callicott, 1980, 321).

Until this point in the discussion, personal ecology has been construed as primarily revolving around the capacity of individuals to create and sustain communities, networks, relationships, and systems that contribute to their self-identity, engender a sense of meaning and belonging, and provide meaningful roles to fulfill. However, with the introduction of an environmental communitarian perspective, emphasis now shifts to a more basic level, as Callicott asserts that all ecological systems, including personal ecological systems, are dependent upon the integrity of the basic biotic community.

Interdependency and Stewardship

Callicott reorients personal ecology. No longer is it sufficient that human beings have a "specific place in the world," it is now necessary for all creatures, and indeed all of nature, to assume their "rightful place" in the world. The land ethic, as articulated by Leopold and Callicott, challenges the core assumptions of the temporary society by insisting that everything and everyone has a role, function, place, and niche, and everything and everyone is interdependent upon one another.

This realization eliminates any pretense one might have that one's personal ecological system, one's "ecology of self" could in any significant way be "self-oriented." Janet Biehl, in *The Politics of Social Ecology* (1998), speaks to the impossibility of being entirely self-oriented when she discriminates between "individuality" and "individualism." According to Biehl, individualism is characterized by "a self-determining individual who, at the supreme moment of his sovereign power, exercises his autonomous will by choosing from among a range of options" (Biehl, 1998, 83). Individualism constitutes the hallmark of liberalism as a political ideology. In contrast, Biehl champions "individuality," which unlike individualism, "gains its very flesh and blood from social interdependence, not from independence, since community support and solidarity provide the context in which the individual acts" (Biehl, 1998, 83).

The process of developing a sustainable personal ecology additionally depends upon promoting this sense of "individuality" via interdependent relationships within the broader biotic community. The pursuit of these interdependent relationships can be achieved in a narrowly utilitarian fashion to promote narrowly conceived self-interests. By comparison, a less self-serving approach to realizing "individuality" can be achieved via the pursuit of ecological stewardship. Stewardship implies a normative expectation that humans have an obligation to serve as caretakers of the Earth's natural communities.

The Self and Other

The sense of duty and obligation that humans may experience for nature is an extension of that which they learn to expect of one another. From a purely psychological and sociological perspective, the isolated, individual, "self" cannot exist in isolation from others. In describing "the ecology of the self," Hormuth observed that "others" play a critically important role, in that they serve to reflect an individual's verbal and nonverbal actions such that the "self" emerges from "the internal organization of external roles of conduct" (Hormuth, 1990, 1).

The critical role that "others" play in the development of self has been recognized for many years. An early philosopher who was instrumental in influencing the field of social psychology, George Herbert Mead (1934), conceived of the self as a "social emergent," derived, in part, by the capacity of individuals to observe one's self and relate to it as an object. From Mead's perspective, self-awareness is developed through the process of encountering "others." Accordingly, Mead writes: "The self is something which has a development; it is not initially there, at birth, but arises in the process of social experience and activity, that is, develops in the given individual as a result of his relations to that process as a whole and to other individuals within that process" (Mead, 1934, 135).

Mead's description of the development of the self, with its emphasis upon the importance of "relations," "processes," and "other individuals," is remarkably similar to Hormuth's definition of "ecology of the self" or personal ecology. Of particular importance to this discussion, however, is the necessary role that the "other" plays in the development of self-identity and in the formation of a personal ecology. One of the great concerns for any writer who contemplates the role and function of a personal ecology is that it will ultimately become self-oriented, insular, and egocentric. Such an orientation would render effective relationships with individuals, communities, and nature unworkable. Indeed, such "self-involvement" is symptomatic of the very forces of alienation and anomie that have been previously discussed. For Mead, the "socially emergent self" approach provides an organic model for human interface with the social environment, and one that parallels the biological relationship between anatomical body parts and the complete human body (Mead, 1934).

Mead's theory of the self as a social emergent serves as the foundation for this book's assertion that environmental and social ecologies are dependent upon the development of functional and sustainable personal ecologies. However, as Hormuth and Mead have observed, self-fulfillment within any personal ecological system requires the presence of the "other," meaning that no complete or functional personal ecology can narrowly focus upon self to the exclusion of "other."

Hormuth (1990) also recognized the important role that environments play in the ecology of self, as they provide a context for individual experience, both self- and other-experiences. In encountering one's self within the context of a particular environment, one learns that one is both self and object, part of the environment yet separate from it. Likewise, in encountering members of the biotic community one may initially perceive that the particular biota is "not self," but, at length, will eventually come to appreciate how each biological constituent is "like self" and will eventually recognize the extent to which one's human self is "biota-like." In each case, people learn to conceptualize their "selves" as being both part of yet different from the biological community. What they cannot rationally do, however, is to conceive of a sense of self in which they do not stand in relationship to the "otherness" of members of the biotic community.

Spirituality and the "Other"

Though everyone learns to understand themselves and their world within the context of the "otherness" of fellow creatures, human beings, and nature itself, some people experience yet another way to relate to the world around them. These individuals relate to the world in a spiritual or religious way and perceive in nature the presence of the divine. Such persons define their separateness and sense of self not simply by comparison and contrast to the "otherness" to be found among other creatures and in nature; they additionally perceive an immanent, creative, and divine "Other."

When considered from a spiritual perspective, natural and biotic communities and ecosystems are construed as having inherent value and worth beyond any valuation imposed upon them by humans. This perspective acknowledges that humanity is dependent upon natural ecosystems for their sustenance and survival, and that the resources found within these natural ecosystems possess inherent and perhaps even divinely imbued value. A spiritually derived sense of stewardship may conceive of the divine as dwelling within nature or the cosmos or as associated with a divine creator. Likewise, such spirituality may be experienced as a natural or supernatural phenomenon or as a product of religious belief or discipline. What matters is that this sense of spirituality, regardless of from where it is derived or how it is experienced, places a normative demand upon the individual to relate to the Earth as a resource to be maintained and protected on the basis of its divinely imbued value.

Such an orientation implies that human beings have a distinct set of rights and obligations toward the natural world. Chief among these rights is the right to consume those natural resources requisite for the survival of humanity and the realization of human fulfillment and satisfaction. Practically speaking, this perspective goes beyond the narrow justification for satisfying human "needs." It additionally justifies the use of natural resources for the realization of human desires and culture. Even so, the realization of these rights and desires is conditional, reflecting the inherent worth of all natural resources and requiring humans to assume a set of environmental management responsibilities that vouchsafe humanity's capacity to exercise the right of consumption — providing such consumption does not threaten the ongoing existence and the integrity of natural resources and systems or dishonor the divine by dishonoring nature.

Admittedly, not everyone would agree that personal ecology must have a spiritual dimension. However, for the spiritually and religiously attuned, there is no question that self-ecologies must be more than merely "other centered" in the sense of affirming the existence and innate rights of other persons, creatures, and environs. They must additionally be "Other centered" in recognition that the world may very well reflect the presence and creative product of the divine.

This, in a nutshell, is the paradox of pursuing a sustainable personal ecology. To be psychologically, socially, politically, or economically sustainable, personal ecologies must be based upon predictable and stable environments that successfully allow people to control their own lives and destinies. These ecological systems must also be grounded in a sense of place, so as to engender self-identity, safety, and a sense of belonging. From a biological perspective, these personal ecologies must vouchsafe the integrity of the broader biotic community if humans are to guarantee their own existence. Moreover, it can also be asserted that to be truly sustainable, these personal ecologies must include a spiritual dimension of one form or another that allows the interests of the individual to be ultimately transcended by those of the natural world, the cosmos, and the divine.

THREE

Social Ecology

···

What Is Social Ecology?

···

Finding a clear, unambiguous definition for the term "social ecology" can be frustrating. The "father" of social ecology, Murray Bookchin, is more prolific at describing what social ecology is opposed to rather than succinctly stating what it favors, or for that matter what it really is. Perhaps his best and clearest definition of the term comes from his initial essay on the topic (Bookchin, 1964) in which he describes social ecology as "an outlook that deals with diversity in an ecological manner — that is, according to an ethics of complementarity."

Although Bookchin indicated in later interviews and essays that he thought he had clearly defined social ecology from the outset, confusion continued to surround the term. Consequently, Bookchin was later asked to clarify what social ecology meant, which he did with the following assertion (Bookchin, 2001, 435): "What literally defines social ecology as 'social' is its recognition of the often overlooked fact that nearly all our present ecological problems arise from deep-seated social problems. Conversely, present ecological problems cannot be clearly understood, much less resolved, without resolutely dealing with problems within society."

Unfortunately, this later effort at clarifying the term still fell short of providing an affirmative definition. Consequently, scholars have been forced to sift through Bookchin's words to detect a definition of social ecology that is more than a statement of what it is not or what it opposes.

Bookchin came closest to making such a definition when he initially described social ecology as an ethic embracing "complementarity." He uses this term to refer to a process in which human beings "complement nonhuman beings with their own capacities to produce a richer, creative, and developmental whole — not as a 'dominant' species but as a supportive one" (Bookchin, 2001, 437). Such complementarity requires engendering a "natural spirituality" among "an awakened humanity to function as moral agents in diminishing needless suffering, engaging in ecological restoration, and fostering an aesthetic appreciation of natural evolution in all its fecundity and diversity" (Bookchin, 1993, 437).

A more complete understanding of social ecology, however, involves appreciating that it "is a coherent form of naturalism that looks to evolution and the biosphere for explanations of natural and social phenomena" (Bookchin, 1994, 236). Moreover, Bookchin describes social ecology as being philosophically organismic, socially revolutionary, politically green, and morally humanistic (Bookchin, 1994, 236–37). These four themes are found throughout Bookchin's work, and there is no mistaking his overarching revolutionary and anarchistic sentiments.

Nevertheless, what most clearly characterizes social ecology is the word "social." This emphasis serves to associate all of society's ecological challenges with long-standing and "deep-seated social problems" (Bookchin, 2001, 435) including "economic, ethnic, cultural, and gender conflicts, among many others, [that] lie at the core of the most serious ecological dislocations we face today" (Bookchin, 2001, 436). Given this perspective, he argues that solving ongoing ecological problems (and by ecological Bookchin means "environmental" problems) necessarily entails successfully resolving society's social problems.

Chief among the societal problems that Bookchin believes must be solved is the problem of Western market economics and its hierarchical underpinnings. In his opinion, Western oriented market-economies pose the greatest threat to the environment. If virtually all of society's ecological challenges can be attributed to social problems, social ecology can be characterized as being as sociological, political, and economic in nature as personal ecology is psychological. Bookchin's solution for remediating this state of affairs entails virtually dismantling Western culture's hierarchical orientation (Bookchin, 2001). Anarchy-oriented social ecologists like Bookchin insist that unless a decentralized society is achieved, one in which vertical, "top-down" hierarchical social and economic relationships are replaced with complementary ones, Western society will witness an environmental collapse on a massive scale (Bookchin, 1982, 1972).

Bookchin's dire prediction is based upon his belief that the "imbalances man has produced in the natural world are caused by the imbalances he has produced in the social world" (Bookchin, 1972, 62). Therefore, he envisions that vouchsafing the future of the natural environment necessitates a dramatic change in the dominant social and economic order.

However, in espousing such radical societal change, Bookchin is careful to distinguish between "ecology" and "environmentalism." Accordingly, he observes that

> environmentalism deals with the serviceability of the human habitat, a passive habitat that people *use,* in short, an assemblage of things called "natural resources" and "urban resources." Taken by themselves, environmental issues require the use of no greater wisdom than the instrumentalist modes of thought and methods that are used by city planners, engineers, physicians, lawyers — and socialists. Ecology, by contrast . . . is an outlook that interprets all interdependencies (social and psychological as well as natural) non-hierarchically. Ecology denies that nature can be interpreted from a hierarchical viewpoint. Moreover, it affirms that diversity and spontaneous development are ends in themselves, to be respected in their own right. Formulated in terms of ecology's "ecosystem approach," this means that each form of life has a unique place in the balance of nature and its removal from the ecosystem could imperil the stability of the whole. (Bookchin, 1980, 270–71)

Given Bookchin's use of the term "complementarity," it would seem that what he opposes in hierarchy is the arbitrary, "top-down" exercise of power and authority within social, political, and economic systems.

Without question, Bookchin's approach to social ecology assumes a radically egalitarian perspective, as well as being notably anarchistic and socially disruptive (Zimmerman et al., 2001). Not surprisingly, given Bookchin's considerable influence, many who have identified themselves with social ecology have also assumed an anarchistic and reactionary posture (Shantz, 2004; Stoll, 2001). Consequently their social ecological philosophies do not lend themselves to the task of fashioning a pragmatic, functional and sustainable approach to social ecology.

For this reason, it is necessary to turn to a critic of Bookchin's approach for a more useful definition of the term. Ecologist John Clark defines social ecology as "a comprehensive holistic conception of the self, society, and nature. It is, indeed, the first ecological philosophy to present a developed approach to all the central issues of theory and practice. It sets out from the basic ecological principle of organic unity in diversity, affirming that the good of the whole can be realized only through the rich individuality and complex interrelationships of the parts. And it applies this fundamental insight to all realms of experience" (Clark, 1990a, 5). This perspective of social ecology places the self and society within the context of a larger whole that encompasses the entirety of the known (and unknown) natural world. Situated within this context, humans find themselves responsible not only for their own fate, but also for the fate of the rest of the natural order.

In this regard, social ecology (as described by Clark, 1990a) is similar to the concept of human ecology as espoused by David Orr (2002), Frederick Steiner (2002), Gerald Marten (2001), and Charles Southwick (1996). While not acknowledged as a "stewardship" approach per se, social ecology, by linking the destiny of the natural world with that of human society, indirectly acknowledges that the fate of the world lies with the choices and actions that humans make as individuals and as societies. In so doing, social ecology eschews Cartesian dualism and seeks to replace this dualistic perspective with what Bookchin refers to as a "dynamic unity of diversity" (Bookchin, 1982, 24). In this formulation, stability is realized not through the pursuit of simplicity and homogeneity but through complexity and variety.

The Importance of Community for Self-Realization

In this chapter I approach social ecology from the perspective of the ecology of the self. In this regard, I share Clark's perspective on social ecology. Central to the concept of social ecology is the importance of the individual self embedded within communities of communities — a self that is at once an "organic whole, yet as a whole in constant process of self-transformation and self transcendence" (Clark, 1990a). Human self-identity is necessarily dependent upon this communal relationship, and in the absence of such a nexus, the self becomes a disembodied entity lacking identity, meaning or purpose.

For most social ecologists, the term community typically implies "commune." Consequently, social ecology understands the self to be communal in nature, reflecting the thought of Martin Buber who asserted that the fate of humankind depended upon their willingness to experience the "rebirth of the commune" (Buber, 1955, 136). Social ecologists believe that the restoration of harmonious relationships among human societies is a prerequisite for achieving harmony with nature. For most social ecologists, achieving harmony necessitates recreating community. John Clark observes that "if human beings cannot develop a deep sense of community, that is, become communal beings, through the actual practice of living within an authentic community of friends and neighbors, then the vast gulfs that separate us from one another can never be bridged" (Clark, 1990a, 10).

This process of becoming communal demands that individuals allow themselves to be transformed via informal and formal interactions with others, and through participation in a myriad of voluntary and democratic relationships and processes. In the absence of such participation and interaction, a "socially competent self" cannot be formed, as the development of such a self is dependent upon "a richly articulated society" (Baugh, 1990, 99). Graham Baugh asserts:

> Without this, the self is reduced to an alienated, monadic ego, morality becomes the mere expression of arbitrary preferences, and reason is reduced to the status of an instrument of achievement . . . If democratic decision-making is to be more than an aggregation of arbitrary preferences, and reason more than an instrument of will, the development of a public "reason" which transcends individual subjectivity, created through interaction with others in a variety of situations and relationships, is absolutely essential. (Baugh, 1990, 99)

Most social ecologists further insist that social intercourse should, to the greatest extent possible, be nonhierarchical in form — based upon the ideal of equal participation. Such a philosophy disallows hierarchical relationships based upon gender, class, social, or economic status or power (Baugh, 1990). At issue, however, is whether it is possible to achieve a functional community that is entirely devoid of hierarchy without resorting to a utopian social philosophy.

Today most approaches to social ecology are articulated from a "community-based" or "communitarian" perspective. The most prominent communitarian social ecologists include John Clark, Daniel Chodorkoff, Amitai Etzioni, J. Baird Callicott, and Wendell Berry. Communitarians seek to promote the development of "communities of communities" and embrace a very different set of philosophical foundations than those espoused by so-called modern liberals or social conservatives. Communitarians generally assume that "modern liberals" perceive that the individual and her/his interests and identity precede any social order, leaving the person free to acknowledge or ignore social influences as a matter of free will (Friedman, 1994). This leaves the individual independent from the demands of social identity or expectations of community. Within this framework, the individual can situate himself or herself (i.e., define his or her "place") as he or she wishes.

This conceptual approach came under attack from a prominent communitarian, Alasdair MacIntyre (1999), who argued that such thinking was a misanthropic interpretation of the original Greek understanding of the importance of community. In MacIntyre's opinion, modern liberalism assumes an excessively individualistic perspective, where the dominance of personal standards renders community standards and values relative, thereby eroding public consensus regarding what constitutes generally acceptable, objective truths and values. For MacIntyre, this elevation of individual values and interests above of those of the society portends a future in which the forces of individual autonomy threaten to overwhelm the forces of social order and equilibrium.

The importance of conceptualizing social ecology from a "community" perspective is difficult to overstate. The community is the context for the development of the ecology of self, and, as George Herbert Mead (1934) has observed, is the place where "self" encounters and is molded by the influence of "others." However, for this process of self-ecological development to be successful and sustainable, it is imperative that the community be as stable and supportive as possible.

To that end, John Clark offers an approach that he refers to as "an authentic" social ecology "inspired by the vision of human communities achieving their fulfillment as an integral part of the larger, self-realizing earth community" (Clark, 1998, 137). According to Clark, social ecology is first and foremost centered around "ecology" or care for the Earth as a "household." Consequently, he argues that humans must relate to the earthly household as household members (householders) and to realize that all issues, problems, and policies are by definition "domestic" in nature. Consequently, in Clark's opinion, "there are no 'non-ecological' social phenomena to consider apart from the ecological ones" (Clark, 2004a).

By comparison the "social" in social ecology is a much more challenging term to comprehend. Says Clark, "There is a seeming paradox in the use of the term 'social' for what is actually a strongly communitarian tradition. Traditionally, the 'social' realm has been counterposed to the 'communal' one, as in Tönnies's famous distinction between society and community, *Gesellschaft* and *Gemeinschaft*." Even so, the complexity of the term's meaning doesn't stop there. According to Clark, fully appreciating the meaning of the term "social" entails remembering that it is derived from the Greek word *socius* or "companion." Consequently, the term "society" implies that there is a relationship between and among a cohort of companions, and the term "social ecology" implies a relationship between and among an even larger and more complex aggregate of animate and inanimate single-organismic and multi-organismic entities and systems. As such, society becomes for Clark merely a "household within the earth household" (Clark, 2004a).

Individuals participate in the social ecology in terms of their individual "self ecologies" — or what I refer to as "personal ecology." Clark's vision of "the ecology of self" affirms the wholeness and completeness of the self but does not perceive it as consistently functioning in harmony with the larger social/environmental milieu. Rather, he conceives of social ecology as consisting of individual "self-ecologies" that are constantly in the dynamic process of "self-transformation" and "self-transcendence." So engaged, the process of defining the ecological self within the context of social ecology is often conflictual and chaotic, experiencing only transient periods of homeostasis — despite the human urge to prolong and maintain such stable and harmonious episodes. From Clark's perspective, the ecological self is "a developing whole, a relative unity-in-diversity, a whole in constant process of self-transformation and self-transcendence" (Clark, 2004b). So conceived, Clark celebrates the multiple facets of self-identity — "the chaos within one" — believing that this characteristic of human nature "attests to the expansiveness of selfhood and to our continuity with the larger context of being, of life, of consciousness, of mind" (Clark, 2004b). More importantly, however, Clark's vision of the relationship between self-ecology and the broader social ecology is one in which "personal self-realization is incomprehensible apart from one's dialectical interaction with other persons, with the community, and with the larger natural world" (Clark, 2004b). In Clark's opinion, the realization of self-identity and meaning is impossible beyond the process of the simultaneous emergence and unfolding of "individuality and social being" (Clark, 2004b).

Daniel Chodorkoff (1990), on the other hand, conceives of social ecology from the perspective of "community development." According to Chodorkoff, community development "must be a holistic process which integrates all facets of a community's life. Social, political, economic, artistic, ethical, and spiritual dimensions must all be seen as part of a whole. They must be made to work together and to reinforce one another. For this reason, the development process must proceed from a self-conscious understanding of their interrelationships" (Chodorkoff, 1990, 71). Chodorkoff's community development philosophy starts with existing communities and extends a vision of what a community could be across a wide array of dimensions. Moreover, and consistent with the notion of building social ecologies upon the foundations of personal ecologies, Chodorkoff's approach to community development begins with the "self" seeking a "self-conscious" understanding of community interrelationships (Chodorkoff, 1990).

In articulating this philosophy, Chodorkoff reflects a perspective earlier popularized by Warren Bennis and Philip Slater in *The Temporary Society* (1968). Considering the fragmented and balkanized nature of modern society, Chodorkoff observes that "the dominant culture has fragmented and isolated social life into distinct realms of experience. The rediscovery of the organic ties between these realms is the starting point for the development process. Once they are recognized, it is possible to create holistic approaches to development that reintegrate all the elements of a community into a cohesive dynamic of cultural change" (Chodorkoff, 1990, 71). Chodorkoff recognizes that social ecology is ultimately concerned with the creation of community rather than the pursuit of narrowly conceived economics. In fact, he observes that community "economic development not rooted in a comprehensive understanding of community may well have a disintegrative effect" (Chodorkoff, 1990, 72).

Such insight, along with his community development orientation to social ecology, demonstrates Chodorkoff's reticence to dramatically disrupt the dominant social order. From his perspective, social ecology "does not propose an abstract ideal society, but rather an evolving process of change, never to be fully realized" (Chodorkoff, 1990, 74). In this manner, he provides a hopeful paradigm for the emergence of a more functional social ecology, while remaining critical of what he describes as a disintegrating modern community suffering from a "culture of domination." Thus Chodorkoff asserts that one of the principal tasks to be accomplished in developing community "is the recreation of local community, and a key component in that task is the identification of humanly scaled boundaries and the reclamation of a sense of place, be it rural village, small town, or urban neighborhood" (Chodorkoff, 1990, 73).

As was the case with the discussion of personal ecology in chapter 2, the concepts of identity (boundaries), belongingness, and sense of place emerge once again. Chodorkoff's vision is similar to that of Wendell Berry (1996) in that he seeks to reinvigorate local communities. Chodorkoff would accomplish this goal by integrating the "cultural traditions, myths, folklore, spiritual beliefs, cosmology, ritual forms, political associations, technical skills, and knowledge of a community" (Chodorkoff, 1990, 72).

Chodorkoff's vision is not without its difficulties. Ideally, the process of developing community should seek to promote a sense of belonging, a sense of place, a set of common experiences and desires, and a sense of shared history and expectations (Allen and Schlereth, 1991). Achieving these goals is difficult even under the best of circumstances, but is particularly challenging in a society that values radical individualism, relatively unfettered competition, ready mobility, and cultural diversity. As Chodorkoff correctly observes, "The growth of values like individuality rooted in community, cooperation, identification with place, and cultural identity are antithetical to the thrust of dominant culture" (Chodorkoff, 1990, 73).

From Chodorkoff's vantage point, both human and natural communities are damaged but not yet destroyed by the forces of modern culture. He perceives that the so-called temporary society has created a vacuum of desire for belongingness that begs to be filled, and one that he is optimistic can and must be satisfied. Central to this vision of community is the restoration of self-reliance among individuals. Accordingly, Chodorkoff asserts that social ecology should promote "the ideal of local self-reliance, and dependence on indigenous resources and talents to the greatest extent possible" (Chodorkoff, 1990, 74–75). In so asserting, he does not foresee that communities could or should ever become completely self-sufficient. Rather he anticipates the development of "communities of communities," or confederations in which interdependent relationships and alliances are forged and sustained. Chodorkoff predicts a degree of political decentralization and the fostering of an ethos promoting sustainable approaches to production, consumption, and waste management.

Chodorkoff's vision is echoed by noted ecologist Wendell Berry in his books *Home Economics* (1987), *Another Turn of the Crank* (1995), and *The Unsettling of America* (1997), among others. One of the central themes throughout Berry's work is his concern over the demise of small towns and rural communities pursuant to realizing the economic and social interests of much larger and more populous industrial communities and economic centers. What results is a glorification of the "large" and "profitable" at the expense of the "small" and "subsistent." Berry summarizes his concerns in *Another Turn of the Crank* by observing that such "economic prejudice against the small has, of course, done immense damage for a long time to small or family sized businesses in city and country alike. But that prejudice has often overlapped with an industrial prejudice against anything rural and against the land itself, and this prejudice has resulted in damages that are not only extensive but also longlasting or permanent" (Berry, 1995, 12). Berry associates the destruction of small-town and rural communities with the destruction of natural environments, observing, "We have much to answer for in our use of this continent from the beginning, but in the last half-century we have added to our desecrations of nature a deliberate destruction of our rural communities" (Berry, 1996, 77).

Berry's prescription for remedying this problem is for individuals to become more self-reliant (as espoused by Daniel Chodorkoff), more grounded in a specific place, locale, and region, more connected with "the land" (particularly by producing some of their own food), and more content to define their individual identities (self-ecologies) and social identities (social ecologies) within the context of a philosophy of mutual self-support and agrarianism. In many ways, Berry articulates a more nostalgic world-view, perhaps to the point of being utopian (van der Leek, 2006). Critics may in fact dismiss Berry's social ecology, claiming that it harkens back to a bygone era that cannot be readily recreated or realized again. However, by formulating his unique vision of "self-ecology" within the context of the importance of "place" and "soil" and employing the power to discriminate between economic wants and needs to challenge the widespread assumption that individuals and communities are helpless in the face of large global economies, Berry provides a powerful, perennially attractive philosophical foundation for social ecology and self-actualization.

Berry's willingness to address economic issues in his social ecology renders his approach particularly useful. Unfortunately, when social ecologists have addressed the issue of economics, they have generally done so in a fashion that calls for societies to function without markets (Staudenmaier, 2003), which, practically speaking, is impossible. Clearly, one of the glaring problems with most approaches to social ecology is their unwillingness to even acknowledge that individuals out of necessity engage in economic behavior that reflects an underlying economic philosophy. This may be due to the lack of recognition that everyone possesses such philosophical values — regardless of whether they are cognizant of them or not — or it may reflect an outright bias toward economics. Regardless of the rationale behind this oversight, an unavoidable facet of all social ecologies is that human beings are by necessity economic beings that must be productive in some fashion within the larger economic system and must adhere to the basic rules and expectations of the larger economic system in the interest of sustaining themselves. In Western societies, that economic philosophy is typically utilitarianism.

Berry, by explicitly calling upon individuals to discriminate between economic "wants" and "needs" and by challenging people to distinguish between "making a living" and "making a profit," brings these underlying economic values to the forefront. In so doing he facilitates the process of considering an approach to self and social ecology that serves to maximize human happiness in a fashion that sustains a quality lifestyle but additionally sustains the social and environmental ecologies upon which such a lifestyle is ultimately dependent. This approach is known as "satisficing" — an economic concept in which the level of economic return sought is deemed to be "good enough" given the values of sustaining social and environmental ecologies, as compared to seeking the maximum return on economic effort or investment (Simon, 1993). By assuming a satisficing approach, Berry makes it possible for individuals to proactively include economic considerations into their formulations of personal and social ecologies. This is an important contribution that no other approach to social ecology so practically and purposefully attempts.

Balancing Personal Autonomy and
Social Order in Communitarianism

One of the major concerns emanating from the thought of social ecologists like Murray Bookchin is the social disruption and chaos that would ensue at the point where radical changes in the current social order were made. Granted, all societies go through periods of relative social upheaval and change, but dramatic and permanent societal change would erode social stability, and the resulting disorder would threaten personal autonomy — thereby thwarting the capacity of individuals to behave in a socially sustainable ecological fashion. In other words, radical societal change would substantially disrupt personal ecology, which would, in turn, necessarily disrupt the social ecology. Since the salient concept related to both personal and social ecology is the word "sustainable" (i.e. the capacity to insure that ecologies persist in a functional and healthy state) it is necessary to develop an approach to social ecology that minimizes dramatic personal and social change.

The communitarian philosophy of Amitai Etzioni is specifically designed to insure that social ecologies develop and thrive in a sustainable fashion — free of unnecessary dramatic social, political, and economic change. Consequently, understanding the principles supporting Etzioni's approach to communitarianism is most useful in terms of articulating a sustainable social ecology.

Etzioni's approach (1998) is designed to seek a continuous balance between the forces of order and autonomy within a society. In so doing, he does not suggest that there is an "ideal" balance between these countervailing forces in any particular society. Instead, he argues that each society is unique, with varying proportions of autonomy and order. In some societies, such as the former Soviet bloc nations, there is an excessive degree of order exhibited within social institutions, whereas in other nations, such as the United States, it could be argued that there is an excess of individualism and autonomy. According to Etzioni, communitarianism seeks to find a balance between these two forces, achieving a new equilibrium gradually and with the minimum social disruption. This approach is unsatisfactory for social conservatives, who seek an even greater degree of social order, as well as for libertarian social ecologists (Biehl, 1998) who seek a dramatic increase in social autonomy by eliminating what they perceive as hierarchical and despotic social institutions.

Those primarily concerned with the ecological welfare of "selves" might be expected to favor an "Etzioni-styled" communitarianism over classical liberalism or social conservatism, because such communitarian thought recognizes the importance of cultural context for the realization of personal identity and seeks to achieve a creative and nurturing balance between the needs of individuals and those of the larger society. It is within this framework that the interface between personal and social ecologies is most likely to be respected and vouchsafed. Having asserted thus, however, there is a significant divergence in opinion to be found among competing visions for ecological preservation and sustainability. These competing visions revolve around very different ideas regarding what constitutes human nature.

Considering Human Nature in Social Ecology: The "Sanguine" and the "Dour"

Etzioni believes that two distinct perspectives regarding human nature dominate philosophy, economics, and public policy — one "sanguine" in character and one "dour." The "sanguine" perspective, which dates back to the Enlightenment, asserts the essential goodness of humanity (Rousseau, 1762) — a sentiment that is often associated with "individualism." This perspective, according to Etzioni, holds "that reason, science, and engineering (including, for some, social engineering) can lead human beings to ever higher levels of sociability if not perfection" (Etzioni, 1998, 161).

A sanguine perspective on human nature, therefore, embraces a shared vision of the essential goodness of humankind but prescribes two very different social remedies to deal with predictable human needs, failings, and excesses. On one hand, it advocates for dramatic decreases in the size and role of government, thereby increasing individual discretion and autonomy. Alternately—and paradoxically—a sanguine perspective may also call for dramatically expanding governmental responsibility toward the individual, seeking to "make good people better" while simultaneously "protecting people from themselves." These complementary outcomes are achieved by significantly increasing government services, and, in so doing, reducing individual autonomy and increasing dependency.

What both approaches to governmental involvement on behalf of citizens have in common is a shared sense of optimism regarding human "progress." This all-important optimism asserts that through reason humans are destined to continually and endlessly improve their condition. Accordingly, "We are not governed by fate, luck, stars, chaos, or random walks, as much less optimistic paradigms assume. Science can banish disease, statesmen and stateswomen can resolve conflicts; policy analysts can fashion effective social programs. The past is dark but the future is light. We are empowered and the world is our oyster, for us to crack open and to enjoy. We are the sun around which the world revolves" (Etzioni, 1998, 163).

The "dour" view, by comparison, characterizes human nature as essentially impulsive, irrational, brutish, and sinful (Russell, 1974, 272). For those embracing this worldview, humankind is at best depraved and at worst evil. This perspective is shared by many Christians, and is consistent with historical interpretations of the doctrine of original sin (Schindler, 2000; Edwards, 1821; Calvin, 1559; Luther, 1525)—a doctrine that has recently been reconsidered in a more optimistic light (Allison and Moore, 1998).

The dour view of human nature is typically embraced by social conservatives who support institutions that exercise significant social control over the populace. Such institutions are expected to educate, indoctrinate, and control people who, in the absence of such oversight, would run amok. This approach to government and social control is also referred to as the "law and order" philosophy. In Western philosophy, this approach, which favors strong government control, is associated with Thomas Hobbes in his famous treatise *Leviathan* (1651). According to Hobbes, humans, in their natural state and in the absence of government and civilized society, would live in a condition of "no arts, no letters, no society, and which is worst of all, continual fear and danger of violent death, and the life of man solitary, poor, nasty, brutish, and short" (Hobbes, 1651). Indeed, so significant was the influence of Hobbes on the shaping of the American character that framers of the U.S. Constitution such as Alexander Hamilton sought to include checks and balances in the American republican form of government that were designed to protect society from the excesses of the free will of the people.

While the sanguine perspective on human nature perceives virtue as residing inherently within human character, the dour school believes that virtue must be ensconced in social institutions. This perspective leads to the belief that morality must be defined and enforced within the community and not at the individual level. Central to the task of community morality is the exercise of one's duty as a citizen, as well as the obligation to maintain strict control over the less acceptable aspects of human nature. Despite the best efforts of each citizen, however, the dour perspective maintains that strong social institutions are required to enforce social order; otherwise individual human inclinations will lead to the utter destruction of the society.

A Communitarian Perspective on Human Nature

Etzioni provides a third perspective of human nature that he refers to as "responsive communitarianism" (Etzioni, 1998). "This communitarian paradigm presumes no superiority of community over the individual (or of the common good over individual rights) but, argues for a social arrangement in which order [ideally voluntary order] is balanced with autonomy, the new golden rule," (Etzioni, 1998, 165). In referring to this "new golden rule," Etzioni alludes to a guiding principle that he proposes for communitarian living which demands that one "respect and uphold society's moral order as you would have society respect and uphold your autonomy" (Etzioni, 1998, xviii).

This communitarian paradigm assumes a dynamic, developmental perspective regarding human nature. It presumes that human nature is neither preordained to be utterly and forever depraved nor completely innocent and good. It does, however, concur with social conservatives in asserting that human nature is somewhat intractable and that people, born as "barbarians," can become virtuous, though never entirely so.

Even so, Etzioni affirms the malleability of human nature by recognizing that human virtue can be realized in the following ways (Etzioni, 1998, 165–66):

- Internalization: through internalizing (as opposed to simple behavioral reinforcement) social values to the point where they became an important part of individual self-identity

- Social formations: through the ongoing development and evolution of social formations (i.e., formal and informal institutions, groups, and organizations) that develop and enshrine social values

- Mediating conflicts and contradictions between forces of autonomy and order: through continual formal and informal dialogue within and between communities, to constantly strive to reduce the tension and apparent contradictions that emerge between the desires of individuals to exercise their autonomy and the needs of the community to promote social order and values

Although these conditions are considered prerequisites for maximizing the possibility for human nature to mature and grow, they are not in and of themselves deemed sufficient to promote a sustainable society.

A Communitarian Infrastructure for Morality and Social Ecology

From Etzioni's communitarian perspective, the process of reliably developing human nature requires a "moral infrastructure," which, for the purpose of this chapter, serves as the core of a basic infrastructure for a sustainable social ecology. According to Etzioni, "The four core elements of the moral infrastructure are arranged like Chinese nesting boxes, one within the other, and in a sociological progression." These core elements include:

- Families that mediate, translate, and transmit human values and shape the "moral self"

- Schools (generically defined) that further contribute to the development of moral character through the additive and corrective interaction of students and instructors

- Communities that serve as the social foundation of the moral voice of the community by drawing upon the resources of religious organizations, voluntary associations, and community institutions and that provide cultural and physical spaces for human interaction, recreation, education, and government
- Communities of communities that serve as the society at large, where moral commitments, values, and sanctions are discussed, mediated, developed, and enforced. Bookchin and others might refer to "communities of communities" as a "confederation" of communities (Etzioni, 1998, 176–77)

This fourfold nested moral infrastructure provides a solid foundation upon which a sustainable social ecology can be developed. Even so, it is lacking in two essential ways. First, the model fails to account for an even more basic level than the family, and one that precedes the first level of social formation — biological identity. Human biological identity precedes social identity and is an essential element for social intercourse. It is also the salient human characteristic that links human beings with the rest of nature.

The second way in which Etzioni's moral infrastructure is lacking is its failure to recognize that "communities of communities" are themselves imbedded in "natural ecosystems" that embrace and incorporate all human culture and community. Just as family social formation is dependent upon the biological identity of its members, so is society dependent upon the larger natural ecosystem for its existence and sustenance. Consequently, a complete, functional, and sustainable social ecology would consist of six levels: (1) biological identity, (2) families, (3) schools, (4) communities, (5) communities of communities, and (6) natural ecosystems. While Etzioni's "nested" orientation to the various levels of ecological infrastructure is maintained, the interrelationships throughout this proposed social ecology become grounded in biological identity and culminate in natural ecosystems. In this regard the social ecology model derived from Etzioni conforms to the "nested hierarchy" paradigm proffered by Hewlett (2003) and utilized as a conceptual model throughout this text.

On one point, all philosophies of social ecology agree. Ultimately, sustainable social ecologies seek to promote the biological and natural ecosystem foundations of human culture and society. Whether they do so for altruistic or utilitarian reasons is largely unimportant, as long as the integrity and future viability of these foundations are insured in a way that respects communities of all kinds. A corollary of this goal requires that factors relating to social stress, disruption, upheaval, and chaos be relieved or eliminated, and that social excesses — whether they involve pollution, overconsumption, population growth, or economic exploitation — be curbed. The various theories of social ecology prescribe different solutions for each one of these problems. I believe that the most pragmatic approach involves the application of communitarian principles — especially those reflecting the economic and "land-based" sensibilities of Wendell Berry and the sensitivity to balancing the needs for individual autonomy and social order that is reflected in the thought of Amitai Etzioni.

While the anarchistic thought of Murray Bookchin will forever be associated with social ecology, I find it to be a dead-end philosophy in terms of its utility for achieving a functional nested ecology that maximizes the capacity for self-ecologies and environmental ecologies to flourish and grow. My disenchantment with Bookchin's ideas is shared by a number of prominent social ecologists (Clark, 1998; Light, 1998; Chasse, 1968) and is best expressed by David Watson, who in his book *Beyond Bookchin* (1996) observes that "Bookchin's once complex, ambiguous ideas have fossilized into dogma" (Watson, 1996, 9). Consequently, I contend that Bookchin and his philosophical kin should be regarded as an evolutionary intellectual tangent devoid of a future. In their place, I would advocate for the development of a renewed, pragmatic, and optimistic social ecology grounded in communitarian principles, dedicated to enabling the best in human spirit and culture to emerge and persist, while doing so in a manner that sustains the biological and environmental ecological resources upon which all life and culture is dependent.

Environmental Ecology

....

What Is Environmental Ecology?

....

Environmental ecology refers to the entire range of natural eco-systems and their constituents—animate and inanimate—residing upon the planet. From this perspective, personal and social ecologies reside within these larger and more complex ecosystems as "communities within communities." They not only owe their existence and sustenance to these natural ecosystems, they are also significant contributors to their functional character and, in the most primary fashion, are little more than a particular biological species and community amidst a plethora of other biotic entities. These larger ecosystems have historically existed without a human presence, and conceivably could do so again. Consequently, the unique human concern regarding the integrity of environmental ecology and the future survival of planetary ecosystems is primarily derived from a concern regarding the future viability of humans and human culture.

This is a subtle but important feature informing most environmental philosophies. Concern over the future of the planet's environmental resources is typically couched in terms of the impact of environmental change upon human beings. In theory, humans could wreak havoc upon the planet's ecosystems to the point where *Homo sapiens* become extinct, and life in some form would likely persist. Predictably, philosophers would decry such an outcome, not just because it would result in human extinction but because other life forms that also arguably had an innate right to exist would be permanently destroyed as well. However, I suspect that the primary issue informing most of those who are concerned about the destruction of natural habitats and ecosystems is the impact that said destruction will have upon human life and culture. After all, people inevitably construe their ecological philosophies from the initial perspective of their personal ecologies and for this reason tend to consider the world from an anthropocentric vantage point.

Such anthropocentric vision is not necessarily a "bad" thing—though it can become a philosophically stifling perspective. Rather than judge the relative value of assuming an anthropocentric perspective upon the world, I would assert that such a perspective is just a fact of human existence and should be accepted at face value. A corollary of this insight is that humans will inevitably and necessarily approach environmental ecology from a utilitarian perspective—even if they simultaneously affirm other values. Practically speaking, everyone has a personal stake in what happens to the world's ecosystems, and it is naïve to assume that anyone can fully articulate an environmental ethic that does not reflect, at least in part, this basic assumption. Consequently, rather than approaching environmental ecology —i.e., the householding of the planet's ecosystems and natural resources— exclusively from a "problem-oriented" perspective, I would advocate that we approach it from the perspective of what serves to "satisfice" personal and social ecological needs and desires while maintaining the integrity and vitality of all those natural ecosystems and their animate and inanimate constituents upon which human life, identity, and culture are dependent.

Beyond an "Ecology of Not": Achieving a Visionary Perspective

I believe that one of the most significant obstacles impeding the development of a sustainable environmental ecology is a penchant among many to primarily focus upon human-induced environmental stress rather than imagining what a sustainable environmental ecology might actually involve. Typical of this perspective is a text by Bill Freedman, *Environmental Ecology* (1994), which describes environmental ecology as the study of the ecological effects of pollution, disturbance, and other stressors. While such study is undoubtedly important, it assumes a remedial perspective and consequently only articulates what environmental ecology "is not"— an "ecology of not"—as opposed to articulating an affirmative vision of environmental ecology.

This chapter seeks to move beyond a mere problem-oriented approach and develop a more visionary perspective of what environmental ecology entails in the fullest sense of the word — in other words, to articulate a vision of what environmental ecology could be. It does so in the interest of providing a lure that can inspire people toward a stronger sense of personal and social obligation for the sustenance and enhancement of natural environments of every type. For a lure to be compelling for action, it must take into account the various perspectives from which people view the environment as well as the myriad uses to which the natural environment is employed.

Ecologist Jean Mercier provides a useful tool for comparing a variety of ecological approaches by creating a theoretical continuum extending between "downstream ecologists" (a.k.a. the "environmentalists"), who are primarily focused upon basic components of the natural environment (i.e. air, water, soil, and bio-ecosystems), and "upstream ecologists," who primarily fall within two groups: (1) the "social ecologists," (anarchists and communitarian-oriented centrists), and (2) the "deep ecologists" (activists who attribute environmental degradation to human encroachment and seek to protect the wilderness by severely curtailing or eliminating human access) (Mercier, 1997). The terms "downstream" and "upstream" reflect the extent to which the philosophical perspective of adherents is focused upon remediation (downstream) or primary prevention (upstream).

Mercier's use of the concepts of upstream and downstream originates from the work of researchers associated with the Natural Step Foundation (Robèrt et al., 1997) — an international organization dedicated to ecological sustainability founded by Swedish physician Karl-Heinrich Robèrt. Robèrt developed what he refers to as "upstream-thinking" as a central feature of "the natural step" (TNS) approach to systematically understanding the underlying mechanisms behind the degradation of environmental habitats. TNS was developed for the explicit purpose of promoting visionary thinking that organizes facts at the micro level with basic principles that define macro-possibilities. This approach has been widely adapted to environmental planning and policy assessment. Paradoxically, upstream thinking has more recently been associated with the work of downstream environmental scientists and conservationists who increasingly seek to place their research and interventions within the context of what Robèrt refers to as "first-order principles."

First-order principles can be thought of as a "trunk and branches," and consequences and activities in the system as "leaves" (Broman, Holmberg, and Robèrt, 2000, 18). Accordingly, "The 'leaves' can be various symptoms that are actually due to neglect of the first-order principles, or measures such as technical designs or changes in behavior as attempts to comply with the first-order principles. Once the 'trunk and branches' are established, decision makers within various fields of expertise can undertake the measures required to meet the principles, 'put on the leaves,' without getting lost at much higher levels of detail than necessary for the decisions" (Broman, Holmberg, and Robèrt, 2000, 18). The value of such a perspective is that it allows environmentalists to develop a vision that is larger and more complete than any particular aspect or feature of the environment that they might otherwise focus upon. By engaging in such upstream-thinking, environmentalists can determine what immediate or intermediate actions relating to sustaining the environment will guide them in the direction of promoting first-order principles, as well as assisting them in avoiding quick-fix solutions and choices that may involve the unnecessary sacrifice of these important overarching, higher-order principles.

Researchers from the Natural Step Foundation define first-order principles as those laws that govern the functioning of the planet. These basic laws of energy and thermodynamics lead to the following conclusions:

"For society to be sustainable, the ecosphere must not be systematically subject to
1. increasing concentrations of substances from the earth's crust.
2. increasing concentrations of substances produced by society.
3. impoverishing physical manipulation or over-harvesting." (Broman, Holmberg, and Robèrt, 2000, 21)

Within the bounds of this definition, the term "systematically" implies that: "(i) the deviation from the natural state must not increase more and more due to the influences from society. (ii) the society must not be organized in such a way that it makes itself more and more dependent on activities that cause such (i) effects" (Broman, Holmberg, and Robèrt, 2000, 21).

These basic tenets become the first-order principles upon which upstream-oriented environmentalists, scientists, entrepreneurs, and other environmentally concerned persons view their interactions with and in the natural environment. These are also the constraints within which social ecology must function to insure that the broader environmental ecology is sustained. From a rational-scientific perspective, this approach is a way to link immediate action to a future vision, even though it is designed around physical constraints existent within the environment. However, these three principles, though necessary, are insufficient for guiding humankind in the direction of achieving the vision of a sustainable environmental ecology. Attention must also be directed toward the human/social element. Accordingly, Broman and his colleagues add a fourth principle: "For society to be sustainable, resources must be used efficiently and fairly to meet basic human needs worldwide" (Broman, Holmberg, and Robèrt, 2000, 22).

This final principle constitutes one of the most significant barriers to achieving a sustainable environmental ecology primarily because of the lack of social consensus regarding what is meant by the terms "efficient," "fair," and "basic human needs" as conceived on a global scale. This fourth principle links a discussion of achieving a sustainable environmental ecology with the preceding discussions of personal and social ecology. In the end, a sustainable environmental ecology has to be one in which humans have a place. This reality challenges the efficacy of the perspective of many deep ecologists, who would principally exclude most humans from what they define as the "natural environment." On the other extreme, many modern industrialists, entrepreneurs, and consumers operate from a perspective in which the natural environment is insufficiently valued, leaving little place for the natural within the built environments of humans.

Deep Ecology

Robèrt's four principles, while essential to achieving a sustainable environmental ecology, are nevertheless couched in the language of "not." So stated, they fail to achieve an affirmative statement of what a sustainable environmental ecology might practically look like. Deep ecology is one upstream philosophical perspective that, while also grounded in the "ecology of not," strives to assert a positive approach to a sustainable environment. The term "deep ecology" (or "deep, long-range ecology movement") was coined by Norwegian philosopher Arne Naess (1989,

1973) and describes an eco-philosophical movement (ecosophy) that is largely grounded in human experience in nature as well as upon the foundation of human knowledge and science. Such deep ecology stands in sharp contrast to so-called shallow ecology that Naess (1973) describes as the "fight against pollution and resource depletion" (97). What Naess refers to as shallow ecology Mercier (1997) calls downstream approaches. However, whereas Mercier characterizes both deep and social ecologies as constituting upstream approaches geared toward prevention, Naess and other deep ecologists discount the value of social ecology, characterizing it as a superficial and narrowly utilitarian ecology with only short-term human interests at heart.

In reality, deep ecology is itself a broad social and ecological movement that includes many variant philosophies beneath its expansive tent, including eco-feminism, proponents of Gaia theory, Native American / indigenous philosophies, Buddhist perspectives and others. While deep ecology began with an emphasis upon what a sustainable environment "was not," it has increasingly become characterized by a growing spiritual perspective of the relationship between humankind and the environment — a perspective particularly influenced by Buddhism and Eastern and Native American / indigenous religions (Simon, 2004; Henning, 2002; Nelson, 1998; Weaver, 1996; Tucker and Grim, 1994; de Silva, 1987).

Regardless of the perspective taken, those identifying themselves as deep ecologists agree that a sustainable environmental ecology must be biocentric or ecocentric rather than anthropocentric in focus. Moreover, it must both respect and celebrate the great diversity to be found within nature. Deep ecologists believe that an anthropocentric approach assumes a worldview in which all of nature exists for human use, while an ecocentric perspective assumes that all organisms and ecosystems are valuable and deserving of protection and sustenance. In this worldview, humans occupy a niche among other creatures in the evolutionary scheme. At issue is just how large and important this niche is.

Deep ecologists have been criticized for advocating a worldview in which humans are viewed as "problematic" in the quest for a sustainable environment. Consequently, protecting the environment necessitates controlling human access or even eliminating humans from threatened environments. Such a philosophy can easily lead to eco-fascism. Deep ecology is often cited as the philosophical foundation supporting radical environmentalism, providing "a diverse body of ideas which taken as a whole express the vision behind the activities" (Manes, 1990, 136). Again, such environmental radicalism more clearly states the "not" of environmental sustainability than articulating a positive image of a sustainable environmental ecology. In fact, radically conceived deep ecology principally articulates the ways in which modern society should be dismantled, making it thoroughly anarchistic in its approach (Manes, 1990).

Radical deep ecologists seek to assert a new spiritual identity that is, as much as possible, divorced from the traditions of Judaism and Christianity. For instance, Gary Snyder (1977, 1975, 1965) and Robert Aitken (1994, 1982) integrate Zen Buddhist spiritual practices into their ecological writing. Meanwhile, *Thinking Like a Mountain: Toward a Council of All Beings,* by John Seed, Joanna Macy, Pat Fleming, and Arne Naess (1988), offers a perspective of humankind existing not above nature, but within it. Consider, if you will, Seed's perspective on humankind's relationship to the earth: "Earth — matter made from rock and soil. It too is pulled by the moon as the magma circulates through the planet heart and roots suck molecules into biology. Earth pours through us, replacing each cell in the body every seven years. Ashes to ashes, dust to dust, we ingest, incorporate and excrete the earth and are made from earth. I am that. You are that" (Seed et al., 1988, 41). It is from this perspective that Seed calls for the convening of what he names a "Council of All Beings" in which humans participate on behalf of all creatures lacking a voice to express their desires and interests. These councils have been convened with increasing regularity as a form of spiritual expression and worship and are developed to experience "mourning, remembering, and speaking from the perspective of other life forms" (Seed et al., 1988, 14).

By excluding human voices and concerns from this spiritual ceremony, the celebrants seek to provide relief from domineering human influence. However, in so doing, they create a spiritual paradigm that excludes human presence and asserts that human beings, as earthly creatures, are unworthy of full participation. Once again, in this spiritual ritual humans are characterized as ill-willed, and a spiritual remedy is enacted that excludes humans from the ritual, just as many other deep ecologists would exclude humans from what they would refer to as pristine wilderness (Canepa, 1997). Such practices reflect a "dour" philosophy regarding human nature.

This dour perspective regarding human nature (Etzioni, 1998) is what fuels deep ecology's intolerance of "anthropocentric" approaches to the environment. For deep ecologists, to be anthropocentric in one's orientation to the world is to believe that all of nature exists exclusively for human use and that humans have the right to exploit nature as a narrowly conceived set of natural resources. This understanding reflects the belief that from an anthropocentric perspective there is an inherent adversarial relationship between humanity and the environment, thereby rationalizing compensatory solutions to this conflict that seek to prioritize the needs of nature over those of human beings.

Imagine if you will, another perspective regarding anthropocentrism in which the term more simply means perceiving the natural world through human eyes and rationality. This is an understanding of anthropocentrism that is much more benign and, I would argue, ultimately inevitable. Such anthropocentrism does not make the automatic assumption that seeing the world through human eyes and conceptualizing its meaning and worth through human culture implies that humans have the right to exploit nature for their own purposes or, for that matter, have the right to subordinate the rights of nature to their every whim and desire. Instead, it simply and pragmatically acknowledges the reality that humans can interact with the environment *only* via human senses and culture and can never in a fully objective way appreciate or realize what it means or "feels like" to be completely biocentric or ecocentric in orientation. At best humans can only appreciate what such orientations might entail if only they were able to fully experience the world through the eyes and senses of other creatures. In this regard, I would assert that far from automatically being considered a "dirty" ecological word, anthropocentrism should be regarded for what it is — an unavoidable reality and an inherent biological human constraint.

However, from within this perspective — a necessary anthropocentrism — there remains wide latitude for behavioral options among humans. Most specifically, this perspective does not preclude discriminating among the relative values of human needs and differentiating between these and the seemingly endless array of human wants or desires. This is the distinguishing characteristic of Wendell Berry's ecological philosophy

(2004) and is that area of legitimate practical and philosophical concern that can be shared across a wide range of ecological philosophies.

Consider if you will the meaning and ramifications of such necessary anthropocentrism. Necessary anthropocentrism implies a human relationship with other human beings and in regard to the entirety of the natural world that is purely a function of human perceptual capabilities and capacity. It further suggests a relationship from which human sustenance, identity, meaning, and purpose are derived via the interface of humans and the external natural environment and everything that it encompasses. Absent the presence of this external natural environment, human existence and identity would be impossible.

While recognizing that necessary anthropocentrism results in a worldview that is uniquely and entirely human-bound in orientation, such anthropocentrism likewise recognizes that all other living creatures also share a capacity for perception in ways that are both enabled and delimited by their unique biological and ecological characteristics. Thus, a necessary anthropocentrism does not assert that a human perceptual framework for nature is the only or even a normative perspective for perceiving and interacting with the natural world. Rather, it acknowledges that all creatures perceive their surroundings in unique ways and in no way suggests that human perception is either better than or, for that matter, similar to the perceptions experienced by other creatures. A necessary anthropocentrism simply acknowledges that human perception is constrained within the limitations of human anatomy, culture, and technology, and that human understanding and behavior is forever tempered and influenced by these inherent perceptual limitations. In other words, being anthropocentric in one's world view is neither good nor bad — *it simply "is."*

Just as necessary anthropocentrism simply is, so is necessary utilitarianism. Necessary utilitarianism logically flows from necessary anthropocentrism, since it recognizes the inherent need of human beings (a need shared by other creatures) to maximize human satisfaction and to minimize discomfort, pain, and misery. Consequently, necessary utilitarianism serves to describe the process by which the human species prospers, grows, and perpetuates itself. So construed, utilitarianism loses its sinister and exploitative character and is instead perceived in its simplest form as the process by which human beings as a species seek to meet their basic needs.

It would appear that human beings share this behavioral trait with a vast array of other creatures — a virtual cross-species commonality.

Within this framework, the natural world with its vast array of creatures and ecosystems ceases to be a treasure trove of resources to be exploited and squandered by "selfish" human beings, but is rather a set of natural resources that humans feel they have the right (by virtue of being fellow living creatures) and an absolute need to utilize and consume. Just as there is no reason for humans to feel guilty for breathing air or drinking water, there is no guilt to be associated with the exercise of such necessary utilitarianism. To be a human creature implies the existential necessity of also being a necessary utilitarian.

As humans are necessarily anthropocentric and utilitarian in their interactions with one another and the natural world, so they are necessarily consumeristic and economic in their behavior and culture. Having culturally evolved well beyond the stage of having to directly and personally derive all resources necessary for human existence themselves, humans rely upon one another to specialize in producing the vast array of resources necessary to sustain life and social culture and acquire these resources from one another via a system of trade or consumerism. This is, pure and simple, another basic characteristic of human beings and their culture. To uncritically demonize markets or market behavior as foreign or antithetical to the survival of species, environments, and ecosystems entails ignoring basic characteristics of human beings and society and engaging in broad generalizations about markets and economies beyond those dictated by reason and nature.

Living within the Bounds of (Human) Nature

Much has been written about man's need to live within the bounds of nature — i.e., within the bounds of the environment — and traits such as anthropocentrism, utilitarianism, and consumerism have been cited as destructive forces that extend human influence beyond natural bounds. By comparison, the ecological literature has failed to appreciate that humans must also live within the bounds of their own human natures, which includes acknowledging that humans are of necessity anthropocentric, utilitarian, and economic both behaviorally and culturally. Consequently, to call upon humans to simply live within "the bounds of nature" (i.e., within the bounds of the environment) alone (Hardin, 1993) is both unrealistic and insufficient to vouchsafe the future of both social and environmental ecologies. Sustainable environmental ecologies are dependent upon sustainable personal and social ecologies, which in turn require that humans live within the bounds of their own (species-specific) nature as well as within the bounds of nature more broadly conceived.

The implications of this basic insight are far-reaching. For instance, recognizing that humans must act within the limits of their own nature as well as within limitations imposed by the broader environment makes it more difficult to simplistically regard environmental issues as the product of human excesses versus nature's needs. In such a formulation, market economics, social and cultural activities, and the pursuit of pleasure and happiness (and the avoidance of pain and discomfort) can be construed as human indulgences purchased at the extent of the natural environment. However, when human nature and culture are introduced into the ecological equation, a very different perspective emerges.

Being human means behaving in a utilitarian fashion, pursuing social and cultural status and meaning, engaging in consumer-oriented economic behavior, and creating a human-oriented built community (or built environment) within the larger natural environment. These human activities are not aberrant, nor are they optional. Rather, these are the things that human beings "naturally" pursue and do. These are the behaviors that humans engage in to realize personal identity and to create and sustain human community in the world. These are not illegitimate pursuits and cannot and should not be discouraged or banned by any ecological ethic or philosophy. Instead, I would suggest that a sustainable ecological approach should recognize and facilitate the realization of human nature at the personal and social level, but do so in a fashion that respects and sustains the larger natural environment upon which personal and social ecologies depend.

One of the guiding tenets internationally used to vouchsafe the integrity of natural environments and to set limits within which human needs and culture can be realized (to include necessary anthropocentrism, utilitarianism, and economic activity) is the "precautionary principle." Simply stated, "The 'precautionary principle' or 'precautionary approach' — is a response to uncertainty, in the face of risks to health or the environment. In general, it involves acting to avoid serious or irreversible potential harm, despite lack of scientific certainty as to the likelihood, magnitude, or causation of that harm" (Precautionary Principle Project, 2003).

This principle serves as a guide for human activity within the bounds of uncertainty and is intended to deter environmentally risky behavior. It does so without prohibiting people from pursuing human interests and needs and without characterizing normal and important human impulses (such as pursuing utilitarian ends or engaging in consumer-oriented economic behavior) as somehow deviant, selfish, or antienvironmental. Instead, the precautionary principle serves as one mechanism (among others) to allow humans to "be human" without imposing a disproportionately excessive "ecological footprint" (Wackernagel and Rees, 1996) upon the environment. Alternately stated, it is a mechanism designed to allow humans to realize their own nature while simultaneously allowing other creatures and ecosystems to realize theirs.

Are "Good Enough" Solutions Good Enough?

This approach is yet another ecological example of what economists refer to as satisficing behavior, in which acceptable outcomes are pursued in the interest of maximizing environmental sustainability values, rather than blithely pursuing any and every imaginable human desire. Satisficing also entails forgoing "ideal" solutions to the competing demands of human and environmental nature in favor of embracing "good enough" (acceptable) options. One can't help but wonder whether good enough is ultimately good enough.

As conceived by Simon (1993), satisficing conceptually incorporates the meaning of the words "satisfying" and "sufficing": an individual confronted with a number of decision options satisfies his or her needs with a choice that is sufficient, though not necessarily ideal. In this approach to decision making, individuals presented with numerous choices select the first available option that they deem to be "reasonable" or "acceptable" rather than pursuing (perhaps indefinitely) the best available (ideal) option.

Critics of applying this decision rubric to environmental issues might question whether satisficing runs the risk of prematurely settling for available and even convenient options when more substantial and sustainable options might present themselves had more discretion, resources, and time been invested in problem solving. Such criticism is absolutely valid. The decision strategies used to solve major environmental problems must appropriately reflect the importance of the issue or problem in question. Obviously routine human decisions do not demand ideal solutions. Neither do consequential issues deserve cursory or narrowly incremental strategies. Satisficing implies that the nature of "good enough" solutions be gauged against the seriousness (importance) of the issue in question, as well as in terms of the decision maker's access to the time, information, and resources needed to make a reasonable informed decision (Wimberley and Morrow, 1981).

Application of the precautionary principle necessarily assumes a satisficing approach to decision making that is tempered against the availability of time, information, and resources. Given these constraints, "good enough" is a relative concept that may prove to be a more adequate decision option under conditions of intense resource availability and considerably less adequate where time, information, and the resources requisite for decision making are lacking. However, the perspective that I am most concerned with is not that of the policymaker or scientist-researcher who is evaluating a substantive environmental issue such as global warming or species extinction. Instead, I am concerned with the situation of the average individual who approaches environmental ecology from a routine, pragmatic, personal, and social ecological perspective.

From Personal Ecological Perspective to Affirmative Ecological Action

One of my primary purposes in writing this book is to articulate an approach that pragmatically describes how an average person interacts ecologically with natural environments. I expect that in most daily decisions individuals will choose to satisfice and will only rarely invest the time and resources necessary to arrive at a more thoughtful and satisfactory solution. Consequently, I would expect that just as humans are necessarily anthropocentric, utilitarian, and economic in their innate behavior, they are likewise necessarily satisficers. Criticizing human beings for what they are not (i.e., not sufficiently deliberate and thoughtful in their choices) does not contribute to achieving a more functional ecological perspective or outcome. However, better understanding and appreciating human nature and typical human ecological behavior can be of great use in motivating individuals to relate to the natural environment in ways that serve to balance an array of personal, social, and environmental needs. In this manner "good enough" solutions become acceptable — not necessarily ideal — and the pursuit of ecological sustainability (at every level) is achieved "one decision at a time," with sometimes better and sometimes poorer decisions and outcomes.

Ultimately an affirmative environmental ecology is one in which humans relate to the natural world upon the foundation of fulfilled personal and social ecologies. In so doing, they approach their relationships with the natural world in sound physical and mental health and with a firm sense of self-identity and place. Furthermore, they enjoy meaningful and stable social roles and responsibilities and are nurtured in every meaningful way by an established and productive society that values their personal autonomy even as they respect and honor the society's mores, values, and boundaries. On the basis of such a sound personal and social foundation, the individual relates to the natural world — the surrounding environments and ecosystems — on a decision-by-decision basis and in so doing seeks to realize personal and social needs and desires while simultaneously demonstrating consideration and respect for the integrity of natural ecosystems and environments — including a regard for the creatures and entities that these ecosystems and environs encompass.

Such considerations happen routinely and daily and are undertaken by ordinary people residing within nested personal, social, and environmental ecologies. There are no "white papers" or research studies developed to evaluate potential priorities and options; no treaties or laws are consulted to guide action. Likewise, there are no consultants or experts called upon to render an opinion or make a decision. Instead everyday people living in their everyday worlds seek to make the best of their daily existence within the confines of their personal and social ecologies, and on the basis of these essential foundational resources interact with the natural world — hopefully devoid of unnecessary want, hunger, ego, or conflict. This is not only an affirmative ecological perspective and vision, it is also a pragmatic real-life approach to sustaining all ecological needs and interests.

The implications of this philosophy for social and environmental ecology are clear and straightforward. I assert that people must first be personally fulfilled and satisfied. This entails much more than the absence of fear, hunger, disease, and hardship. It is an affirmation not only that "basic" human needs are satisfied, it is an affirmation that human imagination, ingenuity, passion, love, and intellect (among other attributes) are also nurtured and sustained. Such satisfaction and fulfillment are in turn dependent upon the realization of a stable, productive, safe, and nurturing social ecology that is more than the sum of the "self-ecologies" it encompasses. It reflects the achievement of a social community and culture that makes human individuality possible, while completely transcending the human potential and worth of any single person. Such a social ecology virtually becomes an independent entity beyond the sum of its constituent parts and from such a perspective creates a climate of order, security, law, culture, ethics, and spirituality that allows individuals to relate to the natural world as something valued and fully "other" than themselves and their culture, while feeling deeply connected to that world, dedicated to its sustenance, and dependent upon its largesse.

A Social Ecological Perspective on an Affirmative Environmental Ecology

The other major objective of this book is to articulate an affirmative environmental ecology from the perspective of social ecology. As I observe above, sustainable social ecologies presuppose that personal ecologies are sustainable and functional. They assume an identity and function that transcends the contributions of individuals and that ultimately becomes normative and historical. Consequently, while social ecology is in large part defined by the patterns of satisficing decisions that individuals make on a regular basis, social ecology cannot be simply defined as the sum of these decisions or as a product or derivative of them. In addition to a pattern of individual actions and decisions that contribute to sustainable social and environmental ecologies, social ecologies must reflect ongoing policies, regulations, and laws that officially and affirmatively commit communities to living within the bounds of natural resources and environments.

To that end, Broman, Holmberg, and Robèrt's (2000, 21) first-order principles must be employed to avoid subjecting the ecosphere to increasing concentrations of substances from the Earth's crust and concentrations of substances produced by society, impoverishing soil and water resources through excessive physical manipulation and over-harvesting, and unfairly impoverishing population groups through the application of inefficient and unfair policies and practices involving the distribution of food and natural resources worldwide. These first-order principles, coupled with the precautionary principle—i.e., "if an action or policy might cause severe or irreversible harm to the public, in the absence of a scientific consensus that harm would not ensue, the burden of proof falls on those who would advocate taking the action" (Hogan, 2007, 8)—can serve as foundations for social ecology considerations that both reflect and moderate the influence of individual ecologically oriented, statisficing decisions.

Such principles do not challenge the "right" of human beings to realize their innate human natures, nor do they challenge the legitimacy of individuals to affiliate and relate in cultural and social ways. However, they do serve to temper human values and behavior—ideally utilizing a consensually derived source of authority that minimally restricts human autonomy and freedom while guaranteeing the current integrity and future sustainability of personal, social, and environmental ecologies. Such principles and practices are based upon a communitarian desire to find a balance between the forces of authority and autonomy while simultaneously seeking to satisfice (not optimize or even harmonize) ecological needs and demands at every level of ecological function.

Assuming that such a process successfully realizes its goals, an affirmative environmental ecology—as opposed to "an ecology of not"—can be realized. Such an environmental ecology ideally envisions a world in which human and nonhuman households coexist within an array of natural ecosystems. Such coexistence reflects the reality that human and nonhuman creatures have an equal right to exist and fulfill their needs by utilizing the natural resources to be found within and upon the Earth. However, since humans possess the capacity for transforming natural environments to a more profound degree than do other terrestrial biota, an affirmative environmental ecology requires that they temper their environmentally oriented behavior by judiciously utilizing the planet's resources to meet their own species-specific needs, minimizing human waste, and pursuing social harmony and affluence in the interest of minimizing social disruption that can ultimately harm natural environs.

Those looking for an elegant and simple vision of environmental ecology will find this vision "messy," incomplete, and unfulfilling. Utopians, like Bookchin-oriented social ecologists and idealistic deep ecologists, will respond derisively to this formulation. However, I believe that those inured in the traditions of environmental stewardship and conservation and those seeking practical and pragmatic solutions to the world's ecological problems will resonate to such an approach. I believe that they will intuitively grasp that effective ecological action presupposes a level of individual and social affluence sufficient to promote sustainable personal and social ecologies and that only after these ecological foundations have been laid can a more encompassing set of social policies, regulations, laws, and treaties be developed and enforced to protect the broader natural environment. Consequently, while this is not an exotic or glamorous solution to the problem of realizing an environmental ecological ethic, it is very much one that I believe serves to translate ethics and values into action and practice.

FIVE

Cosmic Ecology and the Ecology of the Unknown

Human Influence on Cosmic Ecological Systems

The most elementary course in biology acquaints students with the basics regarding how the sun interacts with the climate, soil, and water of Earth to allow for photosynthesis and life as we know it (Martin, 1962). The planet is literally bathed in cosmic radiation not only from the sun, but from the depths of the universe itself. In fact, scientists have known since the late 1990s that cosmic rays produced by distant exploding stars interact with airborne particles in the lower atmosphere to create heavier cloud formations, producing global shading, rain, and cooler temperatures (Marsh and Svensmark, 2000; Perry and Hsu, 2000; Svensmark, 1998; Shindell et al., 1999). Moreover, researchers have discovered that solar winds produced by sunspots can "blow away" sufficient cosmic rays in the atmosphere to reduce cloud cover and influence climate (Svensmark and Calder, 2007; Shaviv and Veizer, 2004; Shindell et al., 2001).

While suspended in the vacuum of the universe, the planet Earth remains inextricably embedded within a cosmic ecosystem that sustains it and largely dictates the conditions under which life on the planet will exist. Given the current state of scientific knowledge, it is now possible to begin understanding how the universe influences life upon the planet. However, unlike in terrestrial ecosystems, it is more difficult to determine what influence human activity has made upon the immediate solar system.

Sadly, humans have spread their litter beyond the limits of the planet. Countless satellite launches, space probes, and spacecraft have left what one 2007 study estimates to be more than eleven thousand detectable objects orbiting Earth (Barker and Matney, 2007; Smirnov, 2001). This orbiting debris poses significant risks to astronauts working on the International Space Station and to spacecraft orbiting the planet and may periodically pose a risk to humans residing on the Earth in the rare instances when large pieces of debris plummet from orbit and strike human dwellings or communities. Thankfully, the U.S. National Aeronautics and Space Administration (NASA) implemented an orbital debris mitigation policy to reduce or minimize the debris field orbiting the planet (NASA, 2007). However, the process of pursuing space exploration and the enhancement of human life via the launching and utilization of orbiting telecommunications and weather satellites has produced a situation in which the integrity of the atmospheric ecosystem that encompasses the planet has been compromised — even if only marginally.

The integrity of our atmospheric ecosystem, which some have referred to as our "near-Earth environment," has become an area of growing concern (Portee and Loftus, 1993), and the U.S. and other nations participating in space exploration have engaged in a process of environmental impact assessment to determine the extent to which human activities have degraded this near-Earth environment (Viikari, 2002). Unfortunately, environmental impact assessments depend upon the capacity of humans to accurately measure environmental influences, and in the case of space environments such activities "cannot necessarily be assessed in detail" (Viikari, 2002, 2). "Outer space" is considered "a global commons and any adverse environmental effects of space exploration and utilization are therefore likely to be severe, irreversible, and expansive in scope" (Milne, 2002; Viikari, 2002). The extent to which humans will feel free to pollute the space surrounding Earth is dependent upon whether space is considered to possess any "intrinsic" value or whether it narrowly serves as a natural resource for human use. Consequently, there exists a degree of uncertainty regarding the ethics involved in polluting space as a matter of principle, as opposed to considerations that narrowly derive from protecting the well-being of space travelers and spacecraft.

Circumterrestrial Space and the Biosphere

Adjacent to the region of "outer space" where spacecraft, satellites, and space stations operate, and within which space debris resides, is an area know as "circumterrestrial space" (Murtazov, 2003). This region includes layers of "high atmosphere" at approximately 200 kilometers, the ionosphere, and the magnetosphere. This "circumterrestrial space" has also been contaminated by humans with a variety of pollutants including fluorocarbons, radiation, ozone, and other chemical compounds. Circumterrestial space is further influenced by accumulations of dust, smog, and other particulates in the atmosphere generated by human technogenic activities. It is also occupied by a variety of agents emanating from outside the planet, including the aforementioned cosmic radiation, protons, electrons, and charged particles. Ultimately this region, bounded by the planet on one extreme and endless space on the other, constitutes an invisible yet vital global ecosystem (Vernadsky, 1991).

Circumterrestrial space and outer space constitute the outermost nether regions of planetary life and are as essential to the sustenance of the Earth as a cell membrane is essential to vouchsafing the integrity of a living organism. Nobody summarized this relationship better than the late Russian scientist and philosopher Vladimir Vernadsky in his monumental work *The Biosphere* (1926). Vernadsky perceived all life on Earth as a unified process extending from the surface of the planet into the biosphere and beyond. In fact, Vernadsky asserts that "life presents an indivisible and indissoluble whole, in which all parts are interconnected both among themselves and with the inert medium of the biosphere" (Vernadsky, 1926, 148).

In so asserting, Vernadsky presents an ecological perspective that unites everything on and around the planet into an all-encompassing and overarching ecosystem that he calls the biosphere, which itself is but a part of a much larger interplanetary and intergalactic system that we might also rightly consider to be yet another ecosystem dimension. However, the hub of this extraplanetary ecosystem is the sun, which is the source of all energy that manifests itself in biospheric life. Accordingly he observes that "*living matter is not an accidental creation.* Solar energy is reflected in it, as in all its terrestrial concentrations" (Vernadksy, 1926, 149).

A Cosmic Ecological Perspective

From Vernadsky's perspective, all known life on the planet occurs on the spherical plane of the Earth — making him the first modern scientist to assert that the planet consists of a self-contained sphere. So terrestrially situated, organic life on Earth creates geology and is not simply the product of geological forces. Terrestrial life has persisted for eons and has steadily incorporated a growing proportion of the Earth's resources into the biosphere (Margulis, 1998, 15). Accordingly, Vernadsky asserts that all life — to include human life — utilizes solar (and we now know interstellar) energy to continuously transform the planet, thereby rendering the Earth as both the "domain and product of life" (Vernadksy, 1926, 149). Thus conceived life is endless and extends throughout space and time as energy and energy forms.

Vernadsky's ideas, while unobtainable to Western scholars during the period of the Cold War, have proven integral to conceptualizing a cosmic ecology. Vernadsky was one of the first in modern times to grasp the significance of what can now be called "cosmic ecology" by recognizing that "Creatures on Earth are the fruit of extended complex processes, and are an essential part of a harmonious cosmic mechanism, in which it is known that fixed laws apply and chance does not exist" (Vernadsky, 1926, 44). Vernadsky's insight into the "cosmic mechanism" that makes all planetary life possible serves to make him an ecological pioneer — perhaps making him the modern era's first "cosmic ecologist." As Lynn Margulis has observed, "What Charles Darwin did for all life through time, Vernadsky did for all life through space. Just as we are all connected in time through evolution to common ancestors, so we are all — through the atmosphere, lithosphere, hydrosphere, and these days even the ionosphere — connected in space. We are tied through an endlessly dynamic picture of Earth as the domain and product of life, to a degree not yet well understood" (Margulis et al., 1998, 18–19).

Vernadsky's vision of the planet and its life forms serves to conceptually integrate the upper reaches of the biosphere — from the planet surface to outer space — with the Earth's terrestrial ecosystems, despite the fact that the biosphere is largely an invisible global ecosystem. So conceived, polluting that ecosystem with gases, radiation, particles, dust, heat, or even space debris no more serves the interest of life on the planet than polluting any other ecosystem — this despite the difficulties involved in measuring human impact upon these nether regions of the atmosphere or impediments to determining their utilitarian value and importance. To adopt a "Vernadskian" ecological perspective is to acquire an ethical sense in which polluting the atmosphere or space is — as a matter of principle — as abhorrent as purposefully despoiling any other terrestrial ecosystem. By extension, logic dictates that human or technogenic pollution of space beyond the confines of planet Earth also constitutes the pollution of a cosmic ecosystem regardless of whether the impact of such pollution significantly impacts these unknown environs or not.

Space Ethics

Such an ethical perspective is much broader and more encompassing than some of the so-called space ethics that have been articulated in recent years. Consider for instance the ethical framework relating to outer space proposed by UNESCO's Commission on the Ethics of Scientific Knowledge and Technology (COMEST, 2000). In developing a set of ethical considerations applicable to outer space, COMEST asserted "the need to regard space as a part of a shared heritage of humankind" such that space be "proclaimed as a scientific territory at the disposal of humankind" (COMEST, 2000, 3).

This ethical assertion completely fails to recognize the inherent right of an ecosystem to maintain its integrity independent of any human utilitarian consideration. Thankfully, COMEST later considered issues relating to environmental ethics and space when they issued a report that at least posed a series of key ethical questions — even if they failed to frame realistic responses — including:

- Should the outer space environment be considered the same way as the terrestrial environment?

- Do we have an ethical obligation to the spatial environment?

- Should we preserve it?

- Does the perspective of space conquest reciprocally modify our perspective on the preservation of Earth? (Pompidou, 2004, 10)

Had Vladimir Vernadsky been afforded the opportunity to respond to these questions, his responses would have undoubtedly been clear and unequivocal:

- Since space is an extension of the terrestrial environment and vice versa, it should clearly be treated the same as terrestrial ecosystems.

- Without question we have an ethical obligation to spatial environments.

- Of course we should preserve these extraterrestrial environments.

- Unavoidably, the process of entering into extraterrestrial spatial ecosystems dramatically alters our understanding of our planet, its resources, and ourselves.

Acquiring a Cosmic Perspective

If we adopt a "Vernadskian" perspective regarding the outer reaches of the Earth's atmosphere, outer space, and the solar system, then we are compelled to relate to these extraterrestrial entities and systems as important, valuable, and life-sustaining. However, if we only relate to those ecosystems immediately adjacent to the human experience in terms of their instrumental value, then humankind may feel justified in regarding outer space, the moon, the solar system, and other planets as "waste sites" where human debris may legitimately be disposed of. Given the narrowly utilitarian state of contemporary space ethics, we are very much at risk of becoming extensive space polluters.

The idea of using space as a repository for unwanted or even dangerous human waste is not a new one. Nuclear power plants, research nuclear reactors, and nuclear weapons plants produce thousands of pounds of high-grade nuclear waste annually, and disposing of such waste has become a major ecological and political issue. Underground disposal of nuclear waste has been the option most favored by regulatory agencies — with the underground caverns beneath Yucca Mountain in Nevada being one of the prime disposal sites. Currently more than 11,000 30- to 80-ton canisters of nuclear waste are buried at this site in more than 160 kilometers of underground tunnels (Coppersmith, 2005). Eventually, this repository is expected to accept a maximum of 77,000 tons of spent nuclear waste, but some more recent studies suggest that, despite geological problems with the site — including a series of underground faults — the waste site may be able to accommodate somewhere between 286,000– 628,000 tons of spent nuclear fuel (Tetreault, 2006).

While most states favor this disposal strategy, residents of Nevada do not, and a growing resistance to further storage is emerging, prompting the search for alternate sites. One solution being proffered — an idea that NASA and the U.S. Department of Energy have been considering since the 1950s and that Boeing Aerospace Company first investigated in 1982 (Boeing, 1982) — is to utilize launch vehicles to propel this unwanted pollutant into outer space at an orbit of 1,100 kilometers or more above the Earth. Proponents of such a solution consider it to be a "safe" alternative for humans and terrestrial ecosystems (Coppersmith, 2005).

In the past, disposing of waste in space was hampered by limitations in the size and capacity of launch vehicles capable of safely propelling the waste out of the atmosphere — requiring multiple and expensive launches to jettison any significant quantity of nuclear debris. Other obstacles also negated space disposal of nuclear waste, including problems associated with negotiating international agreements on how such a program would be managed and regulated (USDOE, 2003). However, with ongoing technological developments in rocketry and space research, who is to say whether practical, political, and technological issues will continue to deter using space as a nuclear dumping ground?

Even so, despite these historical difficulties and the allure of technological space innovation, a central unanswered issue involves whether it is possible and feasible to effectively assess the long-term ecological impact of disposing of highly radioactive waste in an orbit surrounding the planet. Is such a solution truly viable in the long run, and does it not violate — in principle — the value of protecting ecosystems for their own sakes? This issue exemplifies one of the modern pressures that human culture is now capable of exerting upon the upper reaches of the planet's atmosphere and in the surrounding outer space. What this means is that the ecological impact that humans have heretofore exerted on the surface of the planet are now being foisted upon the surrounding cosmic ecology. Who knows what ramifications may follow from using space as a waste site.

If, however, one regards the known universe as a set of interrelated ecosystems that are of great value and importance, then one might relate to what lies beyond the planet's immediate atmosphere with the highest regard and refrain from needlessly polluting it. At issue is whether or not humans derive any sense of meaning or value from these far-flung and ephemeral ecosystems.

Cosmic Purposefulness

John Haught, in his book *The Promise of Nature: Ecology and the Cosmic Purpose,* makes the following observation: "If we believe that the earth is embedded in a meaningful rather than an ultimately pointless universe it cannot help but have a bearing on how we relate to it in our everyday lives" (Haught, 1993, 13–14). Haught's observation relates to the subtitle of his text, *Ecology and the Cosmic Purpose,* and asserts that not only does science suggest that there is a biological and indeed an ecological purpose to the interrelatedness of ecosystems on the planet, but that there is a cosmic purpose reflected in the interrelationships to be found within the universe.

Haught is not alone in this belief in cosmic purposefulness and the need for humanity to come to a clearer understanding of this purpose. For instance, astronomer George Seielstad (1983) has observed:

> Humans occupy a middle vantage point for studying nature. Compared with atoms we are immense; compared with the universe at large we are minute. But ironically, we are made of atoms, and we have descended in a natural way from the universe itself. What is more important, from this middle ground we can reach out, at least intellectually, in both directions to explore the realms of the very small and the very large. Moreover, it is crucial that we do so if we are to understand our role on this planet at this time. (Seielstad, 1983, 2)

Seielstad's belief was also shared by the noted scientist Carl Sagan, who passionately advocated for the exploration of the planetary and stellar systems within which the Earth is embedded. Sagan, like Seielstad, believed that such an exploration was necessary not simply as a matter of human enlightenment but in the interest of breaking free from our "parochial" perspectives of our place on the planet and in the cosmos. To that end he asserts that:

A fundamental area of common interest is the problem of perspective. The exploration of space permits us to see our planet and ourselves in a new light. We are like linguists on an isolated island where only one language is spoken. We can construct general theories of language, but we have only one example to examine. It is unlikely that our understanding of language will have the generality that a mature science of human linguistics requires . . . There are many branches of science where our knowledge is similarly provincial and parochial, restricted to a single example among a vast multitude of possible cases. Only by examining the range of cases available elsewhere can a broad and general science be devised. (Sagan, 1973, 53)

Sagan, Seielstad, and Haught are all concerned that human beings acquire a much broader and deeper perspective regarding their place and purpose in the universe. They are also interested in better understanding the functional relationships between all that exists on the Earth and beyond. However, among these three scholars, only Haught is principally preoccupied with discovering "meaning" in the universe (as opposed to "order" or "function"), and that is because while Sagan and Seielstad look at the universe through the eyes of scientist-astronomers, Haught approaches the issue from the perspective of a theologian. This difference in perspective is important, for it informs how each scholar goes about formulating a solution to the common concern they share regarding the deleterious effect of human culture upon the Earth's fragile ecosystem, and humanity's capacity for potentially despoiling surrounding extraterrestrial ecosystems.

From the perspective of these scholars, broadening and deepening the capacity of humans to understand the interrelationships between this planet and others should serve to reorient them regarding their comparatively unimportant and fragile status within the larger cosmos and promote an appreciation of just how fragile all life is upon this planet and within the universe. Nobody communicates this perspective any clearer than Sagan, who observes that:

The universe is vast and awesome, and for the first time we are becoming a part of it. The planets are no longer wandering lights in the evening sky. For centuries, Man lived in a world that seemed safe and cozy — even tidy. Earth was the cynosure of creation and Man the pinnacle of mortal life. But these quaint and comforting notions have not stood the test of time. We now know that we live on a tiny clod of rock and metal, a planet smaller than some relatively minor features in the clouds of Jupiter and inconsiderable when compared with a modest sunspot . . . No longer does "the world" mean "the universe." We live on one world among an immensity of others. (Sagan, 1973, 51–52)

So informed, scientists hope that humans will hereafter approach their world in a more ecologically considerate fashion and will grasp the importance of protecting the planet and themselves from the excesses of human culture. Consequently, they hope that adopting a cosmic perspective on the universe will enhance mankind's rational capacity for self-preservation and ecological preservation.

The Limits of Rationality and the Ecology of the Unknown

Theologians are particularly aware of the limits of rationality and search for meaning and purpose that exists independently of the human capacity for cognition, ideation, and action. In other words, they conceive of meaning as an independent entity that exists for its own sake and not as a product of human rationality. This is what Haught has in mind when he asks:

> Does the new surge of caringness that many of us now feel toward our planet have the backing of the universe itself, or is our present preoccupation with ecological ethics merely the isolated outcry of a few lonely human subjects marooned on an island to which the rest of the universe is indifferent? Is the universe in its immense expanses of space and time largely unrelated to our moral passion, or is it conceivable that our local environmental concern is in some mysterious way the universe itself crying out for help? Do we inhabit an unconscious universe in which our ethical concern is only an accidental evolutionary anomaly . . . Or is this concern itself the blossoming forth of a cosmic caringness that stems from deep down in the evolutionary process itself? (Haught, 1993, 14)

The theologian's answer to the question of meaning and purpose is one that begins within the realms of rationality but necessarily moves beyond the limits of mind and sensation to embrace faith and a belief in that which is essentially unknown and unknowable.

Scientific knowledge and theory serve as the foundations for formulating a cosmic ecology. Rationality, deductive reasoning, theory building, and hypothesis testing are the tools that are intellectually applied to understanding the place of the Earth, the Earth's environs and humanity within the context of the surrounding cosmos. However, there are limitations to how far or how effectively reason and science can be applied to understanding our cosmic "household" (to coin a phrase applicable to the original Greek meaning of ecology).

Where reason and rationality end a different but related approach to ecology begins. This is the realm of spirituality, theology, ritual, and religion. This demarcation point in understanding our place in the cosmos — where "our" place encompasses all creatures and environs on the planet — is the intellectual territory where John Haught and eco-theologians such as Leonardo Boff (1997), Sallie McFauge (2000), Joseph Sittler (2000), and Matthew Fox (2000) seek to articulate a complementary ecological worldview that extends scientific formulations of our cosmic ecology into the nether regions of "the ecology of the unknown."

Scientific knowledge and the application of reason serve to assuage the human need to know and understand their place in the universe. However, there are inevitably limitations to what can be rationally known at any period in human culture, and historically these limitations in knowledge have not served to constrain human inquisitiveness or concern (Capra, 1975). At every point in time where science and rationality have been exhausted as explanatory tools in understanding humanity's place in the cosmos, theories and belief systems have been developed to extend the human imagination beyond the known world into the unknown. These efforts have largely occurred within the realms of philosophy, spirituality, and religion — although they may also be found within a broad range of fictional literature, poetry, song, and music.

When the human imagination and the desire to find a sense of meaning for existence within the confines of the cosmos is applied to formulating an ecological sense of place and purpose beyond the realm of reason and knowledge, then a new ecological dimension emerges — one that I call "the ecology of the unknown." The "ecology of the unknown" or "spiritual ecology" shares a common feature with all of the other nested ecological dimensions discussed in this text. It is all about "householding" or finding a sense of place, home, and belonging. Just as personal ecology entails the householding task of feeling at home with oneself and within one's immediate surroundings, and social and environmental ecologies involve householding in human and natural families and communities, so does the "ecology of the unknown" or "spiritual ecology" involve feeling at home on Earth, in the universe, in the cosmos, and in the midst of worlds and entities that remain unknown. It is about living with uncertainty in the universe and seeking to find some grounding, orientation, meaning, purpose, and place despite this uncertainty. Alternately stated, "spiritual ecology" — as the personal noetic response to "the ecology of the unknown" is about feeling at home and a part of the creative forces in the universe regardless of whether they entail energy, unified being, a divine immanence, God, or many gods. However, since the ecology of the unknown is manifested in human existence and culture, it will always entail householding — making a home in the world — in the face of what remains unknown and mysterious.

From a conceptual perspective, "the ecology of the unknown" may be dismissed by some as the antiquated reliance upon superstition and religious ritual that humanity has substantially outgrown in this advanced scientific and technological era. It may be considered an anachronism or a primordial and evolutionarily redundant intellectual appendage that is no longer needed or even relevant in our modern world. However, for others, like Haught and his eco-theologian colleagues, theology, religion, and spirituality are not only compelling intellectual and spiritual tools for deriving a sense of meaning and purpose in the universe, they are additionally morally compelling — imposing an obligation upon humanity that emanates from the presence of what is perceived as "the divine" within all of nature and throughout the cosmos. In this sense, spiritually and theologically derived "ecologies of the unknown" become normative structures that not only provide a sense of cosmic place and identity, but additionally provide a sense of cosmic purpose that ultimately serves to define and reinforce a cosmic ecological ethic and obligation to stewardship.

Replacing Ideological Antagonism and
Intolerance with Pragmatism and Tolerance

Unfortunately, divergent views emanating from within the scientific and theological communities regarding the structure, meaning, purpose and importance of the cosmos have increasingly been considered from an either-or and even an adversarial perspective, as if the two views could not coexist together and mutually support human commitment to protecting the environment of the planet and hopefully beyond. Of particular concern is the emergence of a strain of virulent intolerance for the values and practices of religious people as reflected in such recent bestsellers as Sam Harris's *Letters to a Christian Nation: A Challenge to the Faith of America* (2007) and Richard Dawkins' *The God Delusion* (2006) both of which assume a distinctly antagonistic perspective between the worldview of scientists and people of faith.

Of equal concern are the works of Christopher Hitchens, whose *The Missionary Position: The Ideology of Mother Teresa* (1995) betrayed an antagonism toward religion that appears to be even more clearly manifested in his newest book *God Is Not Great: How Religion Poisons Everything* (2007a). While Hitchens' antipathy toward religion and theology is understandable, grounded as it is in the excesses of religious expression and thought, his ultimate stance toward religion is one of remarkable intolerance when he asserts that "all the claims of established religion are bogus, and man-made, and undeserving of anything but contempt and ridicule" (Hitchens, 2007b). Such an assertion is extremely closed-minded, dangerous, and intolerant of religion and the religious.

Arguably, some of this "negative" attitude toward religion can also be found among environmentalists and is in part reflected in the devastating critique of the role of Judeo-Christian religion and the environment presented by Lynn White in his famous essay "Historical Roots of Our Ecological Crisis" (White, 1967). White's critique is still considered to be definitive by many contemporary environmental philosophers and policy analysts. For instance, consider the perspective on religion and the environment articulated by Jason Scorse, an environmental policy professor at the Monterey Institute of International Studies. Scorse characterizes the

God of the Old Testament as a "genocidal terrorist" and dismisses the contents of the Bible:

> After reading these texts I see no reason to associate them with the type of ethic I think is most needed to solve our current environmental problems. In fact, I think the blind faith and fear of science religions bring out in many is exactly the antithesis of what we need most: reasoned debate and an acceptance that we are not as exceptional as we are accustomed to believing. This last point runs entirely counter to the thrust of all religious thought: that we are somehow elevated by our god above all other living things. (Scorse, 2006)

In making this assertion, Scorse engages in some broad generalizations that frankly betray a general unfamiliarity with comparative religion, not to mention alleging religious behavior that cannot be uniformly demonstrated in practice — such that most religions are generally antienvironmental and anti-scientific in their orientations. In making this assertion, Scorse, and others like him, dismiss religion and by extension religious communities and practitioners from any meaningful participation in the current ecological debate. When considered from an ecological perspective, to effectively dismiss from consideration the sizeable contributions that religion and theology (from a number of religious traditions) might make toward redressing the impact that humans are having upon the environment is to effectively disenfranchise a significant portion of the world's population from participating in any meaningful dialogue or planning regarding the future of the planet's ecosystems. In short, it creates a needlessly adversarial, either-or situation.

Ultimately such philosophical "claim staking" is not simply offensive, intolerant, ignorant, and inflammatory, it is also (and I would argue most importantly) counterproductive and non-pragmatic. Such posturing not only fails to address the more important issues to which science and theology are both concerned, it diverts resources and energy away from the central ecological task at hand, that of creating sustainable personal ecologies for people within the milieu of a plethora of human social ecologies, which will not only serve to meet key human needs and desires, but will do so in such a fashion that sustains and vouchsafes the integrity of the Earth's environmental ecology.

I would assert that a necessary prerequisite for formulating such a set of sustainable nested ecologies is the need to locate the entire process of realizing these ecologies (personal, social, and environmental) within the context of the cosmos — a task that Vladimir Vernadsky, Carl Sagan, George Seielstad, John Haught, and numerous other scientists *and* theologians also agree must be realized. In so asserting, I recognize that such a cosmic ecology can be realized from either a scientific perspective (such as that pursued by Sagan, Seielstad, and others) and/or from the perspective of spirituality, religion, and theology (as reflected in the work of John Haught, Thomas Berry, Leonardo Boff, Matthew Fox, and others). Pragmatically speaking, however, I believe that it is necessary to first develop cosmologies that describe what we know about the origins and nature of the universe, as well as what we don't know but believe in or suspect. In short, I believe it is necessary for human beings to articulate a rational-scientific cosmology (regarding the known world) as well as a complementary spiritual-theological cosmology (regarding the unknown).

These complementary cosmologies or worldviews must be compatible with one another to be functional. Spiritual or theological worldviews that are antagonistic to scientific knowledge and rational analysis do not contribute toward developing an ecological perspective that is practical. Moreover, I believe that it is necessary to recognize that people may realistically embrace a number of scientific cosmologies (based upon competing theories and empirical evidence) as well as a variety of spiritual/theological cosmologies. This is to be expected. However, it concerns me whenever either those on the scientific/rational or religious/spiritual side of this issue assert that only *one* perspective is useful or legitimate. Such a formulation guarantees further stalemate on formulating an ecological perspective and ethic and insures that some significant portion of the world's population will be excluded as a potential resource in protecting and managing the Earth's natural resources and environments. I argue, in short, for the realization of a cosmic ecology via the pursuit of cosmology — or the study of the origins and the nature of the universe — from both scientific and theological perspectives.

So stated, I should distinguish between "cosmic ecology" and "cosmology." Cosmology refers to a process of inquiry in pursuit of a world-view that seeks to understand the origins and interrelationships of the universe — interpreting the nature of the universe from either a scientific-rational perspective (Hawking and Mlodinow, 2005; Dodelson, 2003; Hawley and Holcomb, 2003; Hetherington, 1993) or through the prism of religious and spiritual experience or theological doctrine (Forrest, 1996; Davies, 1992, 1982; Polkinghorne, 1986; Whitehead, 1938). Comparatively, "cosmic ecology" refers to our planetary home as a distinct entity — *a place, a household* — situated within the larger cosmos. In this regard, the cosmos serves as the ecological context for the planet, even as the planet Earth provides the ecological context for natural ecosystems, which in turn provide an ecological context for social ecologies that ultimately provide the context for individual human beings and their personal ecologies. In this way "cosmic ecology" is the logical endpoint for a set of nested ecological systems that begin at the personal level and expand to encompass social, environmental, and cosmic proportions. This "cosmic sense of place" represented in cosmic ecology incorporates a perspective on the relationship of all things found within the cosmos — a cosmology — and this cosmology will typically include a scientific-rational perspective as well as a spiritual-theological orientation.

The development of such ecological cosmologies is not simply an important exercise to be undertaken for its own sake, it is one that has been comparatively neglected by environmental philosophers and policy-makers (Skowlinowski, 1990). I believe that one of the reasons for this inattention is that there are a goodly number of competing eco-cosmologies that claim supremacy over one another. Even so, I believe that it is necessary to develop a cosmological perspective or perspectives as a necessary antidote to the modern penchant of relating to the world as a Cartesian dualism in which humans are effectively separated from nature (Grange, 1997, 3). Moreover, developing an ecological cosmology further serves as an antidote to the sense of cosmic homelessness (Haught, 1994) — or in ecological terminology, the absence of a "sense of place" in the universe.

A Two-Dimensional Eco-Cosmology:
Scientific and Spiritual/Religious Approaches

In developing a functional — and, I might add, a pragmatic and practical ecological cosmology — I believe that it is necessary to embrace a two-dimensional approach: a scientific perspective on the nature of the cosmos *as well as* a compatible philosophical, theological, and spiritual orientation. I use the terms "functional" and "practical" purposefully since my rationale for writing this book is to articulate an ecological approach and value system that can be readily appropriated into ecological practice and action. When it comes to developing an eco-cosmology, this is no mean task. By definition, the language and concepts associated with cosmology are often vague, arcane, abstract, and difficult to understand. Given the limitations of such cosmological language, I will endeavor to simplify my rationale for adopting an ecological cosmology.

First and foremost, cosmologies are worldviews (in the broadest sense of the word), where the word "world" is construed in a cosmic sense and not narrowly confined to affairs of the planet. Practically speaking, most everyone possesses a worldview in one form or another. At issue is how inclusive are these worldviews of what we currently know (and don't know) scientifically about the nature and history of the cosmos and our world within it, as well as regarding what we think, feel, and believe in this regard.

The "Standard Theory": Scientific Cosmology

The scientific basis for cosmological ideas has emerged progressively — particularly throughout the nineteenth and twentieth centuries. Within recent years, scientific cosmological thought has been dominated by the so-called big bang theory (now referred to as the "standard theory") first associated with Abbé Georges Lemaître (1933) and later articulated by such scientific luminaries as Steven Weinberg (1977), Joseph Silk (1988), and Denis Sciama (1972), among others. This theory postulates an ever-expanding universe dating back to a specific starting point in time originally associated with the disintegration of a massive primeval atom possessing the atomic weight of the universe (Lemaître, 1927), and more recently associated with the explosive interaction of energy photons known as "ylem" (Alpher and Herman, 1948).

These theories postulate an expanding universe that may (or may not) be limitless in scope. Moreover, recent astronomical investigation and research suggest that the cosmos consists of star systems and galaxies that are constantly in the process of birth, death, and transformation. It would seem that all stars and eventually all celestial bodies enjoy a finite cosmic life span, and that would appear to equally apply to our solar system's sun and to our own planet. As Svensmark and Calder (2007) and other scientists have observed, our planet not only was created by the forces unleashed with this initial "big bang," it is sustained by the cosmic energy released from the disintegration of stars billions of light years distant from Earth. The planet can also be expected to be destroyed by the very forces that created it and sustain it. Logically speaking, at some point in the distant future our own sun will become a supernova, and the planets surrounding it will be destroyed. Consequently, it is possible for humans to conceptualize our cosmic place on Earth as temporary within the greater universe — even though its span of life may encompass billions and billions of years.

A Reverential Universe

Henryk Skowlinowski of the University of Michigan, has long been a proponent of developing an appropriate philosophical approach to cosmology that complements what we scientifically know and don't know about the nature of the universe. He proposes an "eco-cosmology" that seeks to portray the cosmos "in new terms" that emphasize that the "symbiotic, cooperative, just, and equitable structure of the human world is not an aberration, but a natural consequence" (Skowlinowski, 1990, 7) of the human response to the cosmic forces that created the planet, its life forms, and the human species. More specifically, Skowlinowski asserts that it is necessary to "develop a system of values and a concept of the human which coherently fit the image of the *reverential universe* in which we act in a participatory manner" (Skowlinowski, 1990, 10). This image of a "reverential universe" is unique and powerful, suggesting a human attitude and orientation toward the universe that serves as an antidote to what Skowlinowski fears is a dominant tendency to relate to the planet and its cosmic home in arbitrary, relativistic, greedy, slothful, and essentially self-serving ways. Skowlinowski's eco-cosmology emphasizes the majesty and power of the universe as well as the constraints it imposes — but it does so using imagery that is inherently spiritual in nature.

Skowlinowski's spiritual allusion underscores the importance of articulating a spiritual *and* theological cosmology. While there is much that we know and are learning about the nature and scope of the universe from a scientific perspective, there is even more that we don't know and may never know. In other words, our cosmological worldview is dictated by the extent of our knowledge and by the degree of our comparative ignorance. Historically, where science and knowledge fall short and are incomplete, spirituality and religious faith step in to fill the void. Since knowledge is always finite and our rational understanding of any cosmology is limited, I would argue that there is always a need for a spiritual and theological component to cosmology to complement our reason-based cosmic worldview.

Christianity as a Religious Eco-cosmology

In recent years, religious — and particularly Christian — approaches to what Skowlinowski calls "eco-cosmology" have fallen into disrepute — beginning, I think, with the publication of Lynne White's now-famous article (White, 1967), which primarily associates the ecological woes of the modern world with the Judeo-Christian heritage. More recently, John Haught has commented upon the role of religion in contributing to an ecological sensibility and has criticized most religions for their theological emphasis upon the "homeless" condition of human beings and their hunger to find an extraterrestrial, eternal, spiritual home rather than "household" (the original meaning of the word "ecology") here upon Earth. Haught's solution to this problem is to reframe humanity's condition by observing that the real problem is that humankind lives upon a planet, embedded in a solar system and a universe that are also "homeless" and moving toward their ultimate (and perhaps final) destiny in the cosmos. In this way Haught transforms humanity's cosmological perspective from being "lost *in* the cosmos" to being "lost *with* the cosmos," observing that "there is a kinship or togetherness in our mutual forlornness, in our common distance from destiny" (Haught, 1995, 193). Having so reframed the cosmological perspective, Haught makes a bold assertion regarding the orientation of human beings toward the universe:

> What this means theologically is that we can no longer separate concern for our own destiny from that of the whole universe. The cosmos is essentially linked with our humanity. Or better, our humanity is forever situated within the more encompassing framework of a restless universe. And what this means ecologically is that we can no longer plausibly think of the physical universe as though it were not our home. The sense of cosmic homelessness, which underlies so much of our ecological neglect, is no longer intellectually or theologically acceptable. If we ever learn to accept the fact that we do belong to the natural world, something which we have not yet done in a deep way, then we might start treating it better. Cosmology does indeed make a difference to both theology and ecology. (Haught, 1994, 4)

Haught, and others before him, focus upon the emphasis within the world's major religions (Hinduism, Buddhism, Jainism, Judaism, Christianity, and Islam) upon the importance of the afterlife, heaven, or some eternal state of existence over the importance of this earthly abode. His observations that humanity and the universe share a fate and share a home are meant, in part, to counter this otherworldly orientation of many religions and religious people. However, I believe there are ample theological foundations for understanding the planet as not only our home, but a place that humans have a responsibility to and for. Perhaps the salient problem here is the tendency to confuse expectations regarding "paradise" with those of home and to discount the reality of home because of an unrealistic expectation that home once was or should be a paradise.

Such confusion is undoubtedly grounded for Jews and Christians in the Genesis accounts of the Garden of Eden in which the story tells of a divinely created earthly paradise from which man was supposedly evicted, and which resulted in the transformation of the planet from a garden paradise into what it is today. This theme of loss of the garden and the desire to reclaim it is an ongoing one in European and American literature, history, and theology. It was most clearly portrayed by historian Frederick

Jackson Turner in his exposé *The Frontier in American History* (Turner, 1920), in which he characterized the constant pressure to move westward and occupy more and more land as the nation's quest to recapture the essence of the "garden" — the "paradise lost."

What is often overlooked in these theological discourses is that it is paradise that has been lost — *an abstract concept and not an actual place* — not home per se. Home is the world in which humans live and have lived, and it is the site wherein all lived experience occurs. It is an actual location and site. It is a place of birth and death, beauty and ugliness, glory and horror. In these ways, it mirrors the rest of the universe and, frankly — it is what it is. It is not, however, paradise, for paradise is a construct that is situated beyond lived experience. It is an ideal, not reality, and there is much evidence to suggest that a great deal of religious teaching has been directed toward the reality of living in this world as it is — rather than as it is not.

Consider, for instance these words from St. Francis of Assisi's "Canticle of Creatures" (1225):

All praise be yours, my Lord, through all that you have made.
And first my lord Brother Sun, who brings the day . . .
How beautiful is he, how radiant in all his splendor!
Of you, Most High, he bears the likeness.
All praise be yours, my Lord, through Sister Moon and Stars;
In the heavens you have made them, bright and precious and fair.
All praise be yours, my Lord, through Brothers Wind and Air . . .
All praise be yours, my Lord, through Sister water, So useful, lowly,
    precious and fair.
All praise be yours, my Lord, through Brother Fire,
Through whom you brighten up the night . . .
All praise be yours, my Lord, through Sister Earth, our mother,
Who feeds us and produces various fruits
With colored flowers and herbs . . .
Praise and bless my Lord, and give him thanks,
And serve him with great humility.

St. Francis reverentially celebrates the world as he perceives it and senses a divine force that nurtures all life, including human life. His is a celebration of the world and the universe as it is and not a lamentation for a lost paradise.

Similarly, the cosmology of noted Catholic priest and theologian

Pierre Teilhard de Chardin, reflected in his "Mass for the World" from the *Hymn of the Universe* (1961), conceives of human existence as occurring within an actual universe characterized by ambiguity, separation, and incompleteness, yet unified by the presence of an immanent divine force:

> The deeper the level at which one encounters you, Master, the more one realizes the universality of your influence. This is the criterion by which I can judge at each moment how far I have progressed within you. When all the things around me, while preserving their own individual contours, their own special savours, nevertheless appear to me as animated by a single secret spirit and therefore as diffused and intermingled within a single element, infinitely close, infinitely remote; and when, locked within the jealous intimacy of a divine sanctuary, I yet feel myself to be wandering at large in the empyrean of all created beings: then I shall know that I am approaching that central point where the heart of the world is caught in the descending radiance of the heart of God. (de Chardin, 1961, 8)

De Chardin's unique contribution to Christian theology is his assertion that the story and history of the universe and humanity's development, history, and future are inextricably intertwined precisely because of what he perceives as the universality of the divine influence that permeates and unifies both, makes all life possible, and provides a purpose and meaning to existence.

The Anthropic Principle

Interestingly enough, de Chardin's theology anticipated what would later be referred to as the "anthropic principle." This principle was initially proffered by Brandon Carter:

1. We must be prepared to take into account the fact that our location in the universe is necessarily privileged to the extent of being compatible with our existence as observers.

2. The Universe (and hence the fundamental parameters on which it depends) must be such as to admit the creation of observers within it at some stage. (Carter, 1974, 298)

Since Carter's initial formulation of this principle, it has been discussed and elaborated upon by a number of scientists, including Steven Weinberg (1977) and Freeman Dyson (1979). However, the most cogent explanation of the concept was introduced by Heinz Pagels in his now famous "A Cozy Cosmology" lecture (Pagels, 1987). At the outset of this lecture Pagels asserted:

> The universe, it seems, has been finely-tuned for our comfort; its properties appear to be precisely conducive to intelligent life. The force of gravity, for example, could hardly be set at a more ideal level. If it were somehow adjusted upward by just a bit, the stars would consume their hydrogen fuel much more rapidly than they now do. Our sun might burn itself out in less than a billion years (instead of ten billion years), hardly enough time for life as complex as the human species to evolve. If, on the other hand, gravity were nudged downward a notch, the prospects for the evolution of intelligent life would be no less bleak. The sun, now burning more slowly, would cool down and become much too chilly to sustain life as we know it. (Pagels, 1987, 174)

Pagels's assertion suggests a theistic cause behind the ideal environment for life on Earth. The very fact that such a connection can be made has proved to be most controversial for the scientific community. Cosmologist Edward Harrison is credited with being one of the first to address this issue head-on when he observed that the term "theistic principle" could readily be substituted for the "anthropic principle":

> The theistic principle is quite straightforward: the reason the universe seems tailor-made for our existence is that it was tailor-made for our existence; some supreme being created it as a home for intelligent life. Of course, some scientists, believing science and religion mutually exclusive, find this idea unattractive. Faced with questions that do not neatly fit into the framework of science, they are loath to resort to religious explanation; yet their curiosity will not let them leave matters unaddressed. Hence, the anthropic principle. It is the closest that some atheists can get to God. (Harrison, 1981, 157)

Not surprisingly, the anthropic principle — which asserts that humans occupy a preferred place or time in the Universe (Gonzalez and Richards, 2004) — has also been associated with "intelligent design," with speculation as to whether the principle reflects the actions of a divine creator who has purposefully created a cosmos capable of creating life. For instance, Tony Rothman, a physicist at the University of Texas, has observed:

> It's not a big step from the SAP [strong anthropic principle] to the Argument from Design. You know the Argument from Design: it says that the universe was made very precisely, and were it ever so slightly different, man wouldn't be here. Therefore Someone must have made it. Even as I write these words my pen balks, because as a twentieth-century physicist I know that the last step is a leap of faith, not a logical conclusion. Then I reflect: Is it inconceivable that a future civilization will meet God face to face? Will He intentionally reveal Himself? Or will our descendants become God? That is, after all, what the FAP [final anthropic principle] prophesies. In the face of such speculation I retreat and ask: Are the followers of the Anthropic Principle then attracted to it for scientific or religious reasons? . . . When confronted with the order and beauty of the universe and the strange coincidences of nature, it's very tempting to take the leap of faith from science into religion. I am sure many physicists want to. I only wish they would admit it. (Rothman, 1987, 98–99)

Observations regarding the anthropic principle from Rothman, Pagels, and Harrison serve to illustrate why I assert the importance of articulating a scientific *and* a spiritual cosmology.

The Compatibility of Scientific and Spiritual Cosmologies

However, for such cosmologies, to be useful, they must also be compatible. A theological understanding consistent with that of St. Francis, de Chardin, or for that matter a host of other Christian theologians who have proffered eco-cosmologies, complements the anthropic principle quite nicely. Likewise, a theological position embracing "intelligent design" (while also affirming scientific methods and the systematic application of reason and inquiry) might be compatible with the anthropic principle even though some of its critics regard it as an "unscientific" orientation (Drummond, 2005).

By comparison, some theological variations of "creationism" may be partially or even completely intolerant of a scientific perspective on the cosmos (even though it is conceivable that a proponent of intelligent design might simultaneously embrace a scientific *and* a theological perspective). For example the perspectives taken by John Whitcomb and Henry C. Morris in *The Genesis Flood* (1989), or by Morris and Martin E. Clark in *The Bible Has the Answers* (1976) or for that matter the classic book by the creationist William E. Williams, *The Evolution of Man Scientifically Disproved in 50 Arguments* (1928) assume such an antagonistic position toward science and scientific inquiry that their theological cosmologies are essentially incompatible with the dominant "standard" scientific perspective on the origins of the universe. Such perspectives are not conducive to developing an effective eco-cosmology along the lines envisioned within this book.

However, while some evangelical and "fundamentalist" traditions have adopted a creationist worldview that is largely inconsistent with a more scientifically derived cosmology, not all evangelicals and religious conservatives concur. In fact there is a growing evangelical environmental movement that is recommitting to the stewardship expectations implicit in the Genesis creation accounts (particularly the requirement that humankind "till and keep" the Earth [Genesis 2:15]) and in so doing either explicitly embrace ideas consistent with current scientific knowledge and theory, or more indirectly, advocate a cosmological/theological position that does not necessarily preclude accepting a scientific cosmology such as the dominant "standard theory."

This movement is exemplified by the Evangelical Environmental Network that serves as a focal point for much evangelical environmental action. Their efforts revolve around an "Evangelical Declaration on the Care of Creation" which assumes a strong stewardship responsibility, not simply for the environment but also for sustainable and healthy social and ecological environments for women, children, and communities. Among other things, this declaration called upon Christians "to work for responsible public policies which embody the principles of biblical stewardship of creation." The members of this network additionally "recognize that poverty forces people to degrade creation in order to survive; therefore we support the development of just, free economies which empower the poor and create abundance without diminishing creation's bounty" (EEN, 2007).

Yet another evangelical organization committed to Christian environmental stewardship is Care for Creation (CFC, 2007). CFC's board chair, Edward R. Brown, has authored a seminal text for evangelical environmentalists, entitled *Our Father's World: Mobilizing the Church to Care for Creation* (Brown, 2007), that articulates a plan of action for evangelical churches to begin the process of promoting environmental stewardship among their congregants. It has been well received within a number of evangelical congregation and academic settings and has also been endorsed by none other than E. O. Wilson of Harvard University, whose own book, *The Creation: An Appeal to Save Life on Earth* (2007), attempts, from the perspective of an agnostic biologist, to articulate a Christian rationale for protecting the environment.

The passion for reasserting a Christian environmental stewardship ethic among evangelicals is also reflected in books by other noted evangelicals. For instance Tri Robinson and Jason Chatraw's *Saving God's Green Earth: Rediscovering the Church's Responsibility to Environmental Stewardship* (2006), and Matthew Sleeth's *Serve God, Save the Planet: A Christian Call to Action* (2007) both articulate a passionate call to action reminding Christians that their stewardship responsibilities extend beyond a commitment to their families, churches, and communities, but to the world of nature the divine has entrusted them with.

All of these books challenge humans to utilize their resources, economic and scientific ingenuity, and personal assets to vouchsafe the future of the planet, and there is nothing in their ecological worldviews that places them at odds with the dominant scientific cosmology of our time. Moreover, these books encourage Christians to employ their resources to create more sustainable families, communities, and environments, and in this regard appear to espouse religious values and a worldview that is inclusive of the application of science in their daily lives. Their approach is probably more akin to the ecological cosmologies of traditional conservationists relying upon the wise use of science in agriculture and forestry than to many modern environmentalist groups. Indeed, in some respects it may be a throwback to an earlier agrarian era when values from the conservationist movement melded with the dominant biblically informed Christian stewardship ethic that characterized rural communities across the nation during the early to mid-twentieth century.

Islam also approaches the issue of protecting the environment from a stewardship perspective, in part because it shares a number of sacred texts and traditions with Christianity and Judaism. However, while Christians and Jews derive their stewardship authority from the Old Testament of the Bible, and particularly from the two Genesis creation accounts, Muslims derive their authority from the Quran (Sura 33:72), which claims that Allah offered the heavens and the earth the freedom to choose for themselves—a freedom they "refused to bear." However, according to Islam, humans eventually assumed this responsibility, even though they proved to be unreliable stewards given their ignorance and transgressions.

Within Islam, nature possesses a distinct identity and will and exists to praise Allah as do humans. This teaching is reflected in Sura 24:41 which says: "Do you not realize that everyone in the heavens and the earth glorifies Allah, even the birds as they fly in a column? Each knows its prayer and its glorification. Allah is fully aware of everything they do." Unfortunately, the stewardship role bestowed upon humans by Allah may leave some Muslims with the erroneous impression that they are "above" the rest of creation, thereby reflecting an unwillingness to accept the reality that from Allah's perspective, the creation of humankind pales in comparison to the creation of the cosmos (Sura 40:57).

Admittedly, Islamic faith asserts that the earth serves a specific function for humans; for example, "The earth has been created for me as a mosque and as a means of purification" (Al-Bukhari 1:331). However, while the Earth was created to serve the needs of humans, and humans were given a stewardship role to accomplish (vice-regency), Islam does not bestow sovereignty over the world to the faithful. Sovereignty lies with Allah and with Allah alone as is made clear in Sura 24:42–45 of the Quran:

> To Allah belongs the sovereignty of the heavens and the earth, and to
> Allah is the final destiny. Do you not realize that Allah drives the
> clouds, then gathers them together, then piles them on each other, then
> you see the rain coming out of them? He sends down from the sky loads
> of snow to cover whomever He wills, while diverting it from whomever
> He wills. The brightness of the snow almost blinds the eyes. Allah con-
> trols the night and day. This should be a lesson for those who possess

eyes. And Allah created every living creature from water. Some of them walk on their bellies, some walk on two legs, and some walk on four. Allah creates whatever He wills. Allah is Omnipotent.

If one were to imagine two disparate worldviews in which — on one extreme — humans are deemed to be the most important creatures on the planet and by virtue of their elevated status are in charge of all other species and natural resources (the so-called planetary management worldview), or alternately — on the opposite extreme — humans are simply fellow creatures on an equal footing with all other creatures on the planet and with no proprietorship of the planet's living and nonliving resources, then Islam would assume a cosmological position somewhere between these two polar opposites.

Islam conflates these opposing perspectives into a single worldview. Like Christianity and Judaism, it assumes that the divine will is for humans to become stewards of the planet and to care for the creation that Allah has bestowed. However, Islam does not endorse the belief that somehow humans are the most important creature on the planet, nor does it condone the subjugation of the planet or its resources for the selfish purposes of humans (Islam, 2004). While humans have been anointed by Allah to assume the role of "caliph" (*kalipha*) or "vice-regent" over the Earth, they are only authorized to destroy or use (*fasad*) the Earth's resources to satisfy their needs. Authority to use, however, is not authority to abuse the Earth, and humans are required to leave Allah's handiwork (*fitra*) intact out of honor and homage to their divine creator and sovereign (Ouis, 1998; Khalid, 1996).

It must be emphasized that for Muslims, Allah is not simply the Islamic deity for the faithful. Rather, Allah is the lord of all humans and the teachings of Islam are for all humans, with the ultimate goal of making all humans Muslim. However, this religion is *not just* a religion for humans, it is also perceived as a religion for the entirety of creation and as a consequence considers the world, the planets, and the very cosmos to also be Islamic. What renders the Earth (and all of the known universe) Islamic is the "oneness" of Allah with all things and all time — a principle referred to in the faith as *twahid* (unity). Since Allah is at one with all things and all beings, all things and beings are in Allah, thus all things become holy and — by logical extension — Muslim (Khalid, 1996).

This concept of unity (*twahid*) is complemented by *khilifa* (stewardship) and *amana* (trust). For Muslims, stewardship implies servitude to Allah and to the world. However, stewardship is simply a task imposed upon humans. Authority to perform this task emanates from the gift of *amana* or trust from Allah to humans, and religious piety is measured in terms of how well this trust is exercised. While all that exists is one with Allah, and Allah is one with all that is, Allah has assigned a task of caring for the planet to humans (stewardship) and has empowered the faithful to carry out this task with the gift of "trust" (Islam, 2004, 48).

Islam protects the world from deleterious human influence through adherence to an extensive set of teachings relating to cleanliness. The principle emphasis of Islamic ecological practice is cleanliness, particularly in ways that might be associated with modern public health practices. In this regard Islam resembles Judaism. However, this practice occurs within the context of a larger eco-cosmology in which Allah is immanent throughout all that exists, such that polluting any part of the world — including the pollution of one's own body — is to defile Allah and incur his displeasure (Islam, 2004).

Observant Muslims who honor the divine presence of Allah in all of creation are called upon to adhere to *mizan* or the "middle path" (Khalid, 1996). Expectations regarding adherence to this principle can be found in Sura 55:1–12 of the Quran:

The All-Merciful has taught the Qur'an.
He created man
and He taught him the explanation.
The sun and the moon to a reckoning,
and the stars and the trees bow themselves;
and heaven — He raised it up and set the balance.
Transgress not in the balance,
and weigh with justice, and skimp not in the balance.
And earth — He set it down for all beings,
therein fruits and palm-trees with sheaths,
and grain in the blade, and fragrant herbs.
Of which of your Lord's bounties will you and you deny?

This Islamic stewardship mandate is similar to that of Christianity. At issue, however, is whether the Islamic eco-cosmology that informs this stewardship value is compatible with the dominant scientific cosmology (the standard theory). Unfortunately, this question cannot be answered unequivocally. There have been historical periods in Islamic culture in which advances in science and math have flourished as well as those in which they have languished. In modern Islamic culture Muslims have been actively involved in scientific and technological innovation. Islam is not, per se, antagonistic toward science, as observed by the Pakistani physicist and Nobel Prize laureate Abdus Salam (1987) who has reminded those concerned with this issue that approximately an eighth of the content of the Quran requires of Muslims that they seek to discern the presence of Allah in the world around them — an injunction that Salam interprets as a mandate to use science to this end.

Yet this very phenomenon — the use of science to discern and demonstrate the presence of *Allah* in the universe — may also be indicative of the limitations of science to Muslims. Any scientific enterprise that does not have as its principle goal the discernment of Allah in nature (i.e., does not support the interest of a natural theology) may not enjoy religious support (Hoodbhoy, 2007). That Islam might choose to adopt such a fundamentalist perspective on science and scientific research is of great concern to Islamic scientists such as Dr. Salam. He has urged Islamic nations to aggressively pursue scientific research using the following rationale:

> Why am I so passionately advocating our engaging in the enterprise of science and of creating scientific knowledge? This is not just because Allah has endowed us with the urge to know, this is not just because in the conditions of today this knowledge is power and science in application, the major instrument of material progress and meaningful defence; it is also that as self-respecting members of the international world community, we must discharge our responsibility towards and pay back our debt for the benefits we derive from the research stock of contempt for us — unspoken, but certainly there — of those who create knowledge. (Salam, 1987)

Unfortunately, despite Dr. Salam's passionate plea, a realistic assessment of the compatibility of Islamic and western scientific cosmologies must take into account that in an era in which religious fundamentalism characterizes a significant portion of the Islamic world, there is good reason for equivocally affirming the compatibility of the Islamic eco-cosmology with a big bang theory. Nevertheless, just as proponents of "intelligent design" might discern a creator or creative force that predates and precedes the big bang event — and even may have somehow initiated such an event — so it is conceivable that most Muslims may likewise accept a scientifically derived cosmology such as that reflected in the "standard theory" and still perceive the action or presence of Allah behind this initial creative act. What is most significant for the current discussion, however, is that there appears to be nothing inherent in Islamic thought that would necessarily prohibit an Islamic adherent from simultaneously articulating an Islamic *and* a scientific eco-cosmology.

A Confucian Eco-cosmology

Christianity and Islam are not the only religions where examples of spiritual or religious cosmologies can be cited that are compatible with today's so-called standard theory of cosmology. For instance, consider the cosmology associated with Confucian tradition, particularly the philosophy of *ch'i* (Tucker, 1998). The literal meaning of *ch'i* is "material force," implying something embodying matter and energy — concepts that are often categorically segregated in western philosophical thought. *Ch'i* can also be construed to mean a "psychophysical energy," a "vital force" or "vital power," or more diffusely as "that which fills the body" (Tucker, 1998, 190–91).

Unlike Western philosophy, the Confucian principle of *ch'i* does not denigrate matter to a lower ontological order where it can be treated as a mere resource or commodity to be used. Rather, *ch'i* relates to matter and energy as more than mere mechanical or physical phenomena, and accords them a spiritual status and identity. In this way the Western dualism between scientific and spiritual cosmologies is both overcome and unified into a single cosmology that is simultaneously rational, physical, and spiritual.

Ch'i realizes this unity of mind and spirit in a number of ways. First it integrates the emphasis upon immanence found within indigenous religions with the emphasis upon transcendence found in later religious traditions. It also postulates a common life force that permeates all things and unifies the realms of the mind and the soul into a common and reciprocal force, thereby integrating all living things with all the nonliving entities residing within the universe. *Ch'i* also provides a perspective on change and transformation in the universe since it infuses humans and the universe with a common energy, transforms them both, and ultimately explains and justifies the transforming behavior of humans upon the world—a force and energy that humans can actually cultivate and enhance in a moral sense. Rather than seeing humans as mere "observers" of transformation (along the lines of what we encounter with the anthropic principle) *ch'i* empowers and includes human contributions in the transformation of the universe. Consequently, *ch'i* justifies human action and involvement in the world (Tucker, 1998) rather than calling for humans to refrain from acting or to absent themselves from the environment in the fashion that some deep ecologists advocate (Lewis, 1994).

However, the most important characteristic of *ch'i* is the manner in which it unifies humans and nature, as both share a common life force and destiny. Consequently, *ch'i* avoids the dualism prevalent in western philosophies and also serves an egalitarian function by integrating humans across classes and communities, as well as integrating the interests of humans with those of nature. Finally, *ch'i* embodies an empirical orientation that encourages the exploration and understanding of the nature of the interrelatedness of all things (Tucker, 1998). So conceived, this Confucian cosmologic principle is essentially consistent with the dominant scientific standard cosmological theory, particularly in regard to the cosmological vision of Vernadsky (1991) and the empirical findings of Svensmark (1998) and other astronomers and physicists who have documented the myriad influences of life-giving solar and cosmic energy upon the Earth.

Hindu Cosmology and Its Influence upon Buddhism and Jainism

Cosmological concepts inherent to Hinduism, Buddhism, and Jainism have found their way into the mainstream of environmental philosophy, primarily because of their inclusion in the work of the so-called deep ecologists such as George Sessions (1995), Arne Naess (1993), John Seed (1994), and Bill Devall (Devall and Sessions, 1985). Ideas such as the unity of all things, the inherent value of all life and all things, personal responsibility for actions, and the complicated and prolonged sequelae that emanate from every human thought and deed are found throughout these three eastern religions and have been incorporated fully into deep ecology's philosophy. So influenced, deep ecology emphasizes the "universal" character of nature (Scott, 2003) and seeks to protect the environment by encouraging humans to assume responsibility for sustaining its productivity and integrity by limiting their use of and intrusions into natural domains.

Buddhism and Jainism emerged from Hindu traditions. Hinduism, also known as *Sanatana Dharma,* "eternal religion," and *Vaidika Dharma,* "religion of the Vedas," itself emerged from an Aryan (interpreted as "the noble ones") religious tradition grounded in a set of sacred scriptures known as "Vedas." These sacred texts chronicle the evolution of the Hindu tradition from a religion principally focused upon sacrificial practices designed to appease the gods (including Indra, god of the firmament and of thunder and rain, Agni, god of fire, and Soma, god of the moon) to a henotheistic belief in which Brahman, cosmic unity and diversity, and the god Ishvara, the supreme controller, hold dominion over other lesser deities. More specifically, these texts culminate in the emergence of a trinitarian deity encompassing three gods (Brahma, the creator god, Vishnu, the preserver, and Shiva, the destroyer). It is within this composite deity that all exists and finds unity, including the human self known as atman (Narayanan, 2004; Gosling, 2001; Knott, 2000).

Hinduism calls upon its adherents to organize their lives around a set of activities or *purusharthas* encompassing the four aims of Hinduism. These activities include the three goals of the *pravritti* or "those in the world" (*dharma*: religious righteousness [the prime activity]; *artha*: economic success and material prosperity; *kama*: sensual gratification, pleasure, sexual and mental fulfillment); and one goal involving renunciation of the world (the *nivritti*) referred to as *moksa. Moska* entails liberation from the eternal cycle of death and reincarnation based upon past actions (*samsara*) which is considered mankind's supreme goal (Parel, 2006; Schoch, 2006; Coward, 1998; Mumme, 1998; Nelson, 1998; Sherma, 1998; Prime, 1992).

Hinduism, Buddhism, and Jainism all share the concept of the "unity of all things" as well as a belief in *dharma* (meaning eternal order, righteousness, religion, law, and obligation) and *karma* (acts and actions that determine the form in which one is reincarnated). These three concepts render Hinduism and historically subsequent religious traditions ecologically useful since they all share a belief in the inherent interdependency and inter-identity of all living entities with one another, not only from the perspective of a particular entity's lifespan but eternally as each creature is reincarnated as a new life form based upon the consequences of their thoughts and actions. While all three of these religions have served to shape and influence contemporary ecological thought, there are significant differences between them, and not all expressions of Buddhism, Hinduism, and Jainism are equally compatible with what we scientifically know or generally theorize about the origins and shape of the universe. Having briefly reviewed some of the major tenets of Hinduism, let us now turn first to Buddhism and then to Jainism and review their essential beliefs before discussing the cosmological compatibility of these religious traditions with what we scientifically know about the cosmos.

Buddhism: Religion or Philosophy?

Buddhist ideas are prominent in the work of a number of popular ecological writers, including Thomas Berry (2000, 1989), Gary Snyder (1997), Andrew McLaughlin (1993), Joanna Macy (1991), Thomas Weber (1999), and others. Much of the popular interest in Buddhist environmental perspectives has been generated out of a sense of frustration with the culturally dominant Judeo-Christian cosmologies. Moreover, interest in Buddhism has been fueled by the growing cultural gap between the domains of faith and reason. Buddhism has come to be regarded by many as "an 'alternative altar,' a bridge that could reunite the estranged worlds of matter and spirit" (Verhoeven, 2001, 78). More specifically, Buddhism is perceived as being more compatible to science because it neither postulates the existence of a divine presence nor depends upon "God" as a foundation for its cosmology — asserting instead that the universe's function is based upon "natural law" (dharma) — the principle of righteousness, duty, and unity, that which holds the universe together. Compared to Judaism and Christianity, Buddhism presents itself as "a system both moral and physical where everything seemed to work itself out inexorably over vast periods of time without divine intervention" (Verhoeven, 2001, 79).

As Christianity's salience in the eyes of many has waned and Buddhism's perceived relevancy has increased, there has been a tendency to contrast the worst of Christian thought and practice regarding the environment with the most idealistic and "best" of Buddhist thought — thereby undermining any realistic comparison between the two religious worldviews (Williams, 1997, 156–57) and potentially overestimating the potential of Buddhism. Just as there is a tendency to relate to Christianity and Christian cosmology as a monolithic, homogenous worldview, a similar tendency exists regarding Buddhism — a tendency that serves to idealize the tenets of this belief system while ignoring historical and current variations in Buddhist thought (Ingram, 1999).

Buddhist thought has historically evolved across the geographical settings where it has been practiced. For instance, within the rural environs of India, Buddhism related to wilderness areas as dangerous places where faith and human endurance were tested. However, as Buddhism emerged in China, it did so as an urban phenomenon and posited a more idealized and positive relationship between humans and nature (Ingram, 1999). While some characterize Buddhism as being at odds with market capitalism (Sivaraksa, 2002), historically Buddhism emerged and spread as a religion of merchants, and Buddhist thought has always held trade and tradesmen in high esteem (Lancaster, 1997). When Buddhism as an ecological cosmology is discussed, such considerations must be tempered relative to which form of Buddhism is under consideration, the historical orientation of Buddhist beliefs, and in terms of the ecological context within which these ideas are applied. Such considerations are important because Buddhism, like Christianity, exhibits a certain degree of ambiguity in terms of how the human relationship with nature is characterized.

For example, Buddhism has been compared to Christianity and has been praised as an ecologically superior worldview because of its non-hierarchical orientation and its emphasis upon the "natural law" unifying all things (dharma). Dharma (as compared to *nirvana,* which refers to an eternal state beyond suffering) can also be construed to mean "that which sustains the universe" (Yamamoto, 2001; Reat and Perry, 1991) and is often interpreted as the Buddhist idea that most nearly corresponds to the western concept of "nature" (Swearer, 2006; Snyder, 1997).

Despite the aversion of many environmentalists to the concept of hierarchy, Buddhism is unabashedly hierarchical in that it postulates a set of tiered levels of consciousness and being. However, this hierarchical structure does not imply any distinction of privilege or authority, rather describing an ordering conceived around the principles of compassion rather than control. Within Buddhism, nine levels of consciousness are identified, including five that correspond to the various human senses, a sixth "mind consciousness" that integrates the five senses and applies reason, and a seventh "*manas* consciousness" that serves to discern a person's inner spirituality, including perceiving one's sense of self as well as the senses of good, evil, folly, ignorance, arrogance, avarice, and conceit.

In contrast, the eighth level of consciousness is the "*alaya* consciousness," also known as the immortal consciousness, which experiences the processes of birth and death. This level of consciousness is believed to organize and govern the influences of the first seven levels of consciousness. This is the level where all lived experience is captured as "karma seeds" that germinate new forms and levels of existence. This eighth level of consciousness is also referred to as the "store consciousness" since it stores the consequences of all human action. Hierarchically speaking, this eighth level of consciousness is very advanced. However, it is imperfect and can be improved upon at yet a higher level.

The ninth and final level of consciousness is referred to as the "*amala* consciousness," a fundamentally pure level from whence the life force emanates. This is the most advanced state of consciousness and one that remains pure and unstained by any of the acts performed by humans during their lifetimes. To achieve this state of consciousness is to have experienced purity, or pure consciousness. Achieving this state requires one to fully empty oneself of all other earthly content. Consequently, pursuing the ninth level of pure consciousness entails the emptying of all lived experiences in the interest of experiencing pure and eternal consciousness free of all karma and its consequences (Yamamoto, 2001). Achieving such a state would be considered tantamount to realizing nirvana, which according to most Buddhist sects is something that can never be fully achieved, even over several lifetimes (Reat and Perry, 1991).

This idea of ultimately realizing nirvana is closely related to the concept of karma—the theory of moral causation in which actions result in outcomes both within this plane of existence and others. According to Buddhist teachings, existence is not a singular experience. Instead it is an ongoing and evolving process in which people are born and die and go on to live again at another level of existence (i.e., consciousness) with the karma generated by the decisions made in one existence determining what form they will be reborn in. This represents a "what goes around comes around" worldview in which decisions made not only result in consequences in this world, they also relate to how one returns in the next.

Rebirth integrates the fate of humans with that of the rest of nature, since it assumes that every person has existed in some form (i.e., at one of the levels of consciousness) before and will assume a different form of existence in the next life. Rebirth also serves to create a powerful connection between human fate and the fate of other creatures and the world and effectively undercuts the "we-them" dualism prevalent among other western approaches to living with nature. Furthermore it creates a continuity of identity with nature over time and place, since people are believed to experience rebirth over a continuum of time and in different geographical locales. This concept of rebirth that is linked to karma derived from human actions and decisions characterizes not only Buddhism but a number of other eastern religions including Hinduism and Jainism.

Of the various Buddhist persuasions found throughout the world, the form most popular among Western environmentalists is Mahayana Buddhism (Bielefeldt, 2003). Adherents of Mahayana Buddhism are most numerous throughout Eastern Asia and include followers from an array of Buddhist traditions (Warren, 2005). Mahayana Buddhism's popularity among environmentalists is based upon its assertion that there is a unity between all beings and nature (as reflected in karma and rebirth) that stands in sharp relief to the Cartesian dualism characterizing most western perspectives on the natural world. However, at issue is what Buddhism specifically refers to when talking about nature.

It would appear that the concept of "nature" is largely foreign to Buddhism. Instead, Buddhism affirms a hierarchical paradigm of consciousness culminating in a state of pure consciousness that can only be realized over multiple lives, and only then through the complete emptying oneself of all else — present, past, and future. This pursuit of "emptiness" (which is the achievement of the state of spiritual liberation referred to as nirvana) is precisely the idea that the Dalai Lama, perhaps the most visible representative of Buddhism, speaks of when asked to discuss nature. For the Dalai Lama nature implies emptiness, by which he means that emptiness is "the fundamental nature of all reality" (Eckel, 1997).

Such a response does not lend itself to creating a functional eco-cosmology and instead undermines the notion that Buddhism possesses a pragmatic vision and definition of nature. This concern is further exacerbated by the Dalai Lama's claim that achieving unity with nature is not an "expressed aim of Buddhism." Instead, he argues that the principal goal of Buddhism is the purification of the mind (Eckel, 1997). So construed, this approach, rather than being principally oriented to reverence for the natural world (Geraci, 2006), is essentially anthropocentric in its focus and is therefore subject to the same criticism that Lynn White (1967) leveled against Judaism and Christianity years ago.

Buddhism's ambiguous stance regarding nature also applies to its traditional cosmological perspective. As reflected in traditional Indian folklore, Buddhists perceive the universe as being infinite in time and space — filled with innumerable worlds not dissimilar to our own (Reat and Perry, 1991). More specifically, Buddhists perceive two realms of existence located above our known world. The first is "the realm of form" and beyond that "the realm of formlessness." Meanwhile, below the realm of the known world is the "realm of desire," containing six domains encompassing distinctly different categories of beings. Two of those domains include deities (gods and demigods), another encompasses humans, and yet another domain includes animals. Finally, two additional domains are reserved for ghosts and demons (Warren, 2005; Chapple, 1992).

A final aspect of Buddhism that must be considered in understanding Buddhist cosmology is how this tradition relates to creation. This portion of Buddhist cosmology is of particular importance when Buddhism as a religious or spiritual cosmology is compared to the dominant standard theory employed within modern scientific cosmology. Moreover, Buddhism's understanding of creation is also of interest to the extent that it significantly differs from the dominant Judeo-Christian doctrine of cosmology that asserts creation is the product of a single divine presence (God) who created the universe as a single event.

Buddhism departs from Christian creation belief in that it does not include the concept of a "creator" per se (Lande, 2006). Central to Buddhism is the belief that "everything is made from the mind," such that the classically western distinction between subject and object is illusory. So construed, there is no creator beyond the creative power of the mind, and one's perspective upon the world and the universe is completely dependent upon what form one assumes in the world (plant, animal, inorganic substance, etc.). Buddha makes this point in *Avatamsaka Sutra* when he asserts in chapter 20, "The mind is like an artist. It can paint an entire world"; therefore, "if a person knows the workings of the mind, as it universally creates the world, this person then sees the Buddha and understands the Buddha's true and actual nature."

From this conceptual perspective, "nature" is nothing more or less than the mind at work. By comparison, the mind *is everything* since it *conceptualizes everything* (i.e., the entire universe). For Buddhists, the human mind "encompasses the entirety of the universe; there is nothing outside of it, nothing it does not contain" (Verhoeven, 2001, 82). The net effect of such thinking is that the creative force of deity is replaced with the creative force of the mind, rendering any discussions of a "creator" superfluous. This characteristic of Buddhist thought argues that it should not really be considered as a religious perspective (since it posits no all-powerful divinity) but as a philosophical, mystical-spiritual orientation.

The Buddhist "post-creator" perspective has attracted the interest of some of the world's best thinkers such as Albert Einstein, Alfred North Whitehead, and Bertrand Russell. These intellectuals and others have hailed Buddhism as "the religion of the future," a heartening example of applied metaphysics on a grand scale, and a promising amalgamation of speculative and scientific philosophy that "takes up where science cannot lead" (Verhoeven, 2001, 83). This optimism regarding the compatibility of Buddhism and modern science was crystallized in the 1970s by Fritjof Capra in *The Tao of Physics* (1975), which served as an intellectual catalyst for considering science and religion/spirituality within the shared perspective of systems theory. Capra's ideas are born out of his recognition that all rational approaches to reality are limited. This assertion was revolutionary in that it ran counter to the era of scientific optimism that had dominated the world, particularly since the beginning of the "Sputnik era" in which scientific innovation and knowledge seemed to be expansive and endless. However, Capra challenged this historical optimism regarding the scope of the boundaries of scientific innovation and knowledge by essentially asserting that rationality can only carry one so far, and beyond that limit, something more was needed.

Granted, this argument of the limitation of science was not a new one (indeed it is a distinction prominently asserted in this text). For instance, religious communities have made this argument for some time as a rationale for conversion. What made Capra's position unique is that it posited a creative force in the human mind that heretofore had not been widely asserted, and it placed all beings and things within the context of an "organic whole in which all parts are interdependent; a dynamic system which is self-balancing and self-adjusting" (Capra, 1979, 5, 7). It is from this perspective of a dynamic system operating on its own set of "natural laws" that Capra admonishes scientists to expand their horizons and approach their work within the context of "larger systems" — by which he means considering all science within the context of physical, social, cultural, and cosmic dimensions — thereby "transcending present disciplinary boundaries and expanding their basic concepts from their narrow, reductionist, connotations to a broad social and ecological context" (Capra, 1979, 7–8).

Stripped of the presence of a divine creator, and freed from the institutional orthodoxies that religions typically impose, Buddhism as a philosophical and spiritual cosmology — especially as envisioned by Capra — is clearly compatible with scientific ideas about the creation of the universe. Similarly, it constitutes a worldview that is compatible with viewing existence from the perspective of systems theory and ecology (Sponberg, 1994).

Buddhism as a religious orientation, however, is a more complicated affair. Since modern Buddhism does not affirm a creator, some would say that, strictly speaking, it is not a religion. One might argue, however, that since it is directed toward a transcendent idea or goal — in this case toward nirvana or the achievement of complete peace, transcendence, and the elimination of all suffering — it does indeed constitute a religion. Likewise, since religions typically encompass a set of rituals, mythologies, and sacred texts, Buddhism could most definitely be considered a religion.

However, as a religion, it includes many characteristics (such as the multiple levels of existence encompassing gods and demigods and demons and devils) that are mythological and incompatible with a scientific worldview. Similarly, its emphasis upon human fulfillment and self-realization and its insistence upon the creative power of the mind may be considered to be excessively anthropocentric in orientation — a criticism that has been widely directed toward Christianity. Moreover, there appears to be significant variation in the practice of Buddhism across the Asian continent and worldwide that reflects divergent local geographic and cultural factors as well as variations in the evolution of Buddhist thought over time. Therefore, while it is possible to unequivocally affirm the basic compatibility of

Buddhism as a philosophy with the dominant standard big bang scientific cosmology, it is not possible to as completely assert the same degree of compatibility for Buddhism as a religion.

Within western societies, Buddhism is primarily expressed as a spiritual orientation and as a philosophy if not a psychological orientation to living. In fact Buddhism appears to occupy a significant niche in the religious and spiritual life of the West as a religion without God — an idea that has gained growing legitimacy in an increasingly scientific, reason-oriented, secular world. Given the current form and presence of Buddhism as a philosophy and value system in the West, there is every reason to be optimistic about it as a spiritual cosmological complement to modern scientific cosmology. However, when one considers the problems attendant to Buddhism (its anthropocentric orientation, its hierarchical organization, and the mythological components of the faith), its extensive following worldwide (estimated by the Pluralism Project in 2007 to range from 2.5 to 4 million) and the degree of diversity with which this faith is expressed globally, there is reason to temper our estimation of the extent to which Buddhism as a religion can serve (worldwide) as a compatible religious and spiritual complement to the dominant scientific cosmology of our time.

Jainism and Non-violence toward All Living Creatures

Like Buddhism, Jainism is an ancient religious tradition that dates back to 500 BC in India. However, unlike Buddhism, which spread well beyond the bounds of India, Jainism remains comparatively isolated to India. Jainism has not received as much emphasis within the environmental community as Buddhism, but its influence has been reflected in the work of a number of deep ecologists, including Michael Tobias (2000), deep ecology's poet laureate, Gary Snyder (1974), ecologist and ethologist Marc Bekoff (2007), and the originator of "deep vegetarianism," animal-rights advocate Michael Allen Fox (1999).

Jains seek to live an existence that is completely peaceful and harmless to any other creature, no matter how large or minute. This is, of course, a very difficult lifestyle to sustain, and no adherent of this faith is ever completely successful in not harming any creature. In pursuit of this goal, most Jains are vegetarian, but even adherents to this dietary regime ultimately find themselves consuming some creature living in or upon a vegetable or fruit. The demands of Jainism prohibit adherents from engaging in agricultural occupations. Consequently, most Jains work in professions such as business, banking, technology, medicine, engineering, and the like (Eck, 1997).

Like Buddhists and Hindus, Jains seek a life of discipline that will propel them from one "life existence" to another in search of what they call *moksah,* a term also common to Hinduism — a religion with which many adherents of Jainism identify themselves (Flugel, 2006) — which quite literally means being liberated from the cycle of life, death, and rebirth in which all creatures are embedded. Essentially to achieve *moksah* is to realize nirvana. For Jains, achieving this goal is realized through acquiring a lifestyle in which every effort is made to not harm any living creature — a philosophy known as *ahimsa.*

Jainism's cosmology is described in their holy text, the *Tattvartha Sutra* — meaning "that which is" (Tatiia, 1994, 33) — and is defined by a universe filled with innumerable living souls, or *jivas,* who are repeatedly reincarnated into a new life form based upon their experiences and actions in earlier existences. Many of these existences (as many as seven or eight consecutively) can be in the form of a human being, but other forms of being (*gati*) include "hell dwellers" (*naraki*), plants and animals (designated with the single term *tiryanc*), and deities (*dev*). This cosmology emanates from earlier Aryan traditions and is very similar to that espoused within Hinduism. However, a unique feature of Jain cosmology is the manner in which animate and inanimate nature are mutually deemed to possess life, including earth-bodied, water-bodied, air-bodied, and fire-bodied entities, vegetation experiencing only the sensation of touch, and an array of mobile beings exhibiting from two to five senses that may or may not be sentient (Tatiia, 2002, 3).

Jain cosmology is further distinguished relative to the hierarchical position of life forms based upon their capacity for sensation — which ranges from sensation and touch among lower life forms like worms, to smell, sight, and the combination of all these sensations with the capacity for hearing among higher-order creatures (Chapple, 2006, 149). Within the bounds of this cosmology, human beings represent the most developed of all life forms but are in no way privileged relative to other living creatures. Paradoxically, the relationship of human beings to all other living creatures is achieved by following the principle of "being yourself" — albeit doing so in recognition that all other living entities are indistinguishable from oneself such that the destruction of any living thing constitutes an act of self-destruction. Consequently, the world of humans is one of singularity in which there is no absolute distinction between "self" and "other" (or "subject" and "object") as is found within a Cartesian worldview (Plumwood, 1994). Humans and all other living entities share the common identity of being alive and in community with one another, creating a state of affairs in which "being yourself" as a human being occurs in context and consideration of the equal rights that the entire living community enjoys to existence and self-realization (Bhaskar, 2002). Within the confines of this cosmological paradigm, "being yourself" demands pursuing the realization of nirvana by "giving yourself" to other living entities "in such a way that there is no self and no other" (Kumar, 2005, 64).

Harm directed toward any living creature results in what the Jains regard as a thickening of the "veil of karma" that surrounds every person. The thicker this veil becomes the less vibrancy the soul exudes and the dimmer the life force of the person. Consequently, the goal of Jainism is to minimize the thickness of this veil and to avoid those forms of karma that are believed to inhibit the realization of *moksah* (Tatiia, 1994). These accretions can only be removed by fervent, daily discipline encompassed in a set of eleven principles that followers dedicate themselves to incorporating into their daily lives, including nonviolence (*ahimsa*), truthfulness (*satya*), refraining from theft (*asteya*), sexual restraint (*brahmacarya*), non-acquisition (*aparigraha*), physical labor (*shariashram*), healthy nutrition (*aswada*), courage (*sarvatra bhaya varjana*), religious tolerance (*sarva dharma samantava*), local economy (*swadeshi*), and respect for all beings (*sparsha bhvana*) (Kumar, 1999b, 285–304). The extent to which followers embrace these principles depends upon the degree of devotion to the faith they choose to exhibit, with the most devout adherents represented among the Jain priesthood and lesser degrees of adherence demonstrated among the laity. However, the obvious connections between Jainism values toward living creatures renders all followers of this faith environmentalists to a level that even devout deep ecologists might emulate.

The most recognized Jain ecologist is philosopher Satish Kumar, who achieved notoriety as the editor of the popular bimonthly ecology and spirituality journal *Resurgence*. Kumar's goal has been to make a connection between the spiritual and ecological dimension of human beings, to achieve what he refers to as "reverential ecology" (a concept similar to Skowlinowski's 1990 vision of a "reverential universe") or the capacity to discern the sacred to be found within all creatures and all things (Kumar, 1999a). Kumar presents this as an alternative to what he sees as the western penchant to adhere to a more narrowly utilitarian ecology that tends to prioritize human wants and needs over those of all other creatures.

One of Kumar's most significant contributions to the realization of "reverential ecology" is his paradigm for integrating the eleven sacred Jain principles with ecological sensibility (see Kumar's 1999 book *Path without Destination*). So construed, Kumar provides a blueprint for achieving a Jainist ecological perspective built around:

- Nonviolence (*ahimsa*) — refraining from contributing to the destruction of nature or engendering disharmony,

- Truthfulness (*satya*) — refraining from manipulating people or nature,

- Refraining from theft (*asteya*) — restraining from taking more from nature than we absolutely need,

- Sexual restraint (*brahmacarya*) — acknowledging the integrity of the human body, society, and nature and not exploiting these natural entities beyond what is needed to achieve self-sustenance and fecundity,

- Non-acquisition (*aparigraha*) — committing to liberate oneself from non-essential possessions, thereby reducing consumption and natural resource utilization,

- Physical toil (*shariashram*) — engaging in physical labor as a means toward desiring fewer possessions, ultimately reducing the use of natural resources,

- Healthy nutrition (*aswada*) — pursuing a largely vegetarian existence that harms the fewest possible life forms,

- Fearlessness or courage (*sarvatra bhaya varjana*) — trusting the natural world to care for and fulfill needs and desires even as humans seek to nurture other life entities,

- Religious tolerance (*sarva dharma samantava*) — allowing for the divine to be perceived in the natural world in different ways by different people and recognizing the divine in nature,

- Local self-sufficiency (*swadeshi*) — realizing a sense of place in the world and being content within this ecological space, and

- Respect for all beings (*sparsha bhvana*) — recognizing the unity of existence that all life shares and demonstrating self-respect by respecting all other life. (Kumar, 1999b, 285–306)

Reverential ecology operates upon the principle that all natural things are holy — possessing the divine, not as a transcendent presence but rather as an immanent one. So construed, all of the Earth's life forms and resources constitute not only a source of sustenance and personal identity, they also represent a divine obligation to relate to all within the world as an integral expression and extension of self. In this the Jain expectation that you "be yourself" as a unity of "self" and "other" is interpreted as a divine mandate to additionally "care for yourself" — which unavoidably mandates caring for all other living entities (Kumar, 2005, 76).

Religious Responses to the Ecology of the Unknown

If there is a common theme to be found within the religious traditions of Confucianism, Hinduism, Buddhism, and Jainism, it is the emphasis upon the unity of all existence and all things such that the fate of human beings and that of the rest of the animate and inanimate universe are fully intertwined. Little wonder that these ancient religious traditions have become so popular among those concerned over the fate of the planet's ecosystems and creatures. Moreover, these religious traditions may also be incorporated into the lives and habits of humans as either a religious practice and ritual or more informally embraced as a philosophy of living. In this way the central values of these eastern religions have imbued an otherwise secular orientation toward protecting and preserving the environment.

In fact, for many, the attractiveness of the ecological values that emanate from these religions is precisely the degree to which they allow for the expression of a spiritual orientation toward the ecology of the unknown without requiring adherence to a religious discipline or membership in a religious group or congregation. In short, they allow for the expression of a set of eco-friendly spiritual values that can be incorporated into the lives of individuals without demanding any other moral obligations beyond those individuals impose upon themselves. Finally, these values (the unity of all creatures and creation, the immanence of the divine in all things, the imbuing of all matter and all creatures with life-giving energy) are fully compatible with affirming a rational, scientific cosmology along the lines of those associated with the "standard theory." At the risk of oversimplification, I would assert that the values enshrined in Confucianism, Hinduism, Buddhism, and Jainism are those that are most readily associated with the modern environmental movement—particularly among deep ecologists.

On the other hand, Judaism and Christianity suffer by comparison as spiritual responses to "the ecology of the unknown"—as would Islam if its stewardship orientation toward the environment were better known and appreciated. More specifically, they tend to be discounted as excessively anthropocentric and devoted to a deity that is more transcendent to the world of nature than immanent within it. I further suspect that in an era in which secular orientations toward the environment are extremely prominent within the ecological literature, the theological demands and obligations of Islam, Judaism, and Christianity are considered to be excessively demanding, authoritarian, and at odds with the desire by many to further enhance individual autonomy and independence (Dunlap, 2004). Moreover, the Judeo-Christian tradition is interpreted as providing the philosophical/theological foundations for the Enlightenment which has since influenced western capitalism and the scientific revolution. In short, in the opinion of many left-oriented post-Enlightenment environmentalists, these religious traditions constitute the core foundations upon which environmental destruction and devolution have been rationalized (Bourdeau, 2004; Geisinger, 1999; White, 1967)—this despite the fact that such an assertion has been widely challenged and discredited (Minteer, 2005; McGrath, 2002; Kearns, 1996; Moncrief, 1970).

Consider, for instance, the critique of the relationship between Judeo-Christian tradition and modern science proffered by Meera Nanda, a fellow at the John Templeton Foundation and author of *The Wrongs of the Religious Right* (2005), and *Prophets Facing Backward* (2003):

The epistemological assumption is that due to the combined sins of the Judeo-Christian heritage, the "violence" of the Scientific Revolution, patriarchy and capitalism, Western science is a reductionist and dualist science which separates matter and mind and treats nature as a mere object to be studied by a pure subject removed from it. The holy grail of the anti-Enlightenment left is to create a postmodern science which is non-dualist and holistic. Such a science will be more respectful of the embeddedness of science and nature itself in the cultural context. Non-Western people, especially the oppressed among them, are supposed to provide the right kind of metaphysic and the right kind of non-dualistic, non-reductionist value orientation to study nature. (Nanda, 2004, 3)

This critique clearly articulates a very popular sentiment among anti-Enlightenment environmentalists and suggests why the eastern religious traditions have been identified as potential sources of "nonindividualistic" and "nonreductionistic" ecological values.

However, as Nanda goes on to suggest, these eastern values are most acceptable when they are articulated at a "philosophical" level. When they are espoused as religious values demanding religious practice and devotion, then they become less acceptable. In fact, they may be perceived to be every bit as toxic and deleterious to the environment as Judeo-Christian values when they are cloaked in the language and expectations of religious dogma and practice.

A prime example of this can be found in India, where a Hindu "dharmic ecology" has emerged that integrates ecological initiatives with religious practices and piety. Meera Nanda's description of this uniquely Indian religious ecology is illustrative of this phenomenon (quoted material to follow includes language definitions from *The Oxford Hindi-English Dictionary* [McGregor, 1997] and from *The Hindi-English/English-Hindi Dictionary and Phrasebook* [Scudiere, 2003]):

In order to mobilize the masses, the mainstream Indian environmentalists have not shied away from invoking Hindu imagery and myths. Just about every popular Hindu ritual or idea has been tapped for its potential for mobilization on behalf of the environment. Examples range form women tying *rachis* [leaf stems] to trees, mass recitations of *bhagwat purana* [a religious narrative on realizing a state of well-being] at the site of *Chipko* [referring to an Indian environmental movement protecting trees from loggers by hugging them or attaching themselves to the trunks], fasts, religious vows on the river banks and temples, invocations of *Krishna* as the lord of cows and pastures, invocations of *shakti* [sacred force], *devi* [the great goddess of Hindu tradition], *bhu mata* [mother earth] (or *Narmada mata* [mother who gives pleasure, or mother of the river Narmada], or *Ganga mata* [mother goddess or mother of the river Ganges]), *karma* [fate, destiny], reincarnation, sacred trees, rivers, and even *jati* [birth], reinterpreted as biological species living in harmony with their environment. *All* major environmental campaigns in recent years, including *Chipko, Narmada Bachao Andolan* [Save the River Narmada Movement], and even to some extent, the controversy over the GM [genetically manufactured] seeds, have had their share of religious imagery, mixed in with the nostalgic invocations of the good old days.

While some might welcome this marriage of Hindu faith with ecological sensibility, Nanda, who describes herself as a secular environmentalist, is highly critical of this union of Hinduism and ecology. She is particularly critical of what she describes as "dharmic ecology," which she characterizes as

a curious third position — a hybrid position, as postcolonial theorists would have it — between the traditional left and the traditional right: *they have radical left-wing politics, but a radical right-wing epistemology and cosmology.* That is to say, they claim to bear allegiance to the traditional left's political ideals of equality, peace, tolerance and ecological sustainability, but they have lost faith in the traditional left's cultural ideals of scientific reason, naturalism, humanism and secularism. They reject these Enlightenment ideas as the source of colonization of the non-Western societies, and a cause of the environmental and other problems facing the modern world. (Nanda, 2004, 2)

Nanda asserts that advocates of this ecological ethic share a set of core assumptions with what she describes as the "anti-Enlightenment left" that include the following:

- because Hindus find gods in nature, because they see nature as embodiment of the divine, they must therefore, by definition have a more evolved ecological ethic; and

- it is because of the colonization of the mind by Western reductionist science that Indians have forgotten this holistic worldview; and

- that a revival of this "holistic" "non-dualistic" worldview is needed in order to encourage environmentally responsible development. (Nanda, 2004, 4)

Despite its promise of restoring traditional values and relationships with nature, Nanda finds reason to fear the emergence of such dharmic ecology, believing that it is profoundly at odds with the prevailing rational, reason-based scientific cosmology such as that which has been discussed throughout this chapter.

Dharmic ecology is, from Nanda's perspective, an admixture of left-oriented political ideology and right-wing, pre-Enlightenment religious tradition. She characterizes this ecological orientation as being a "Trojan horse for *Hindutva*" (literally meaning "Hinduness"; referring to Hindu nationalism movements) that serves as a nexus for paganism internationally conceived (Nanda, 2004, 4). As such, it is an orientation that introduces Hindu religion into politics by way of an environmental ethic. Dharmic ecology, Nanda asserts, is a particularly attractive ethic for anti-Christian and neo-pagan groups across Northern Europe — groups she characterizes as consisting of "disillusioned Christians" who find themselves drawn to paganism because it embraces an immanent vision of divinity rather than the comparably transcendent God of their Judeo-Christian tradition (Nanda, 2004, 6).

What concerns Nanda about dharmic ecology's popularity among neo-pagans outside of India is that she fears (a) this ecological approach will promote the spread of neo-paganism, which may contribute to the growth of fascist and racist factions that are sometimes attracted to neo-paganism, and (b) the spread of neo-paganism will further erode the influence of Christianity allowing for religious Hinduism to spread internationally, which will in turn enhance the authority and influence of Hinduism within India (Nanda, 2004, 7).

Given Nanda's second concern, it would appear as though she is particularly troubled by Hinduism being propagated as a religious faith by way of an ecological ethic. This is a similar reservation that has been expressed regarding evangelical Christianity's interest in environmentalism (Beisner, 1997). Nanda challenges what she believes to be the central rationale among environmentalists for embracing this ecological approach—namely that "a religious attitude of sacredness and reverence toward nature encourages wise use of Nature" (Nanda, 2004, 7). Instead she advocates for a secular environmental approach that separates the sacred—as a religious belief—from science.

Nanda's critique is of interest not only because she so articulately addresses what she perceives to be the negative ramifications of adopting a religious approach to environmentalism (regardless of whether the religion is Hinduism, Christianity, Judaism, or for that matter—I would imagine—any religion), but additionally because she suggests that, despite the fondest desires of deep ecologists and others who have infused eastern religious values into their ecological ethics, it may not be possible to integrate religious ideas into an ecological ethic without religion utilizing the "ethic" as a vehicle for achieving its primary goal of attracting adherents to the faith. In short, she challenges one of the central ideas of this text, namely that spirituality and religion can indeed be utilized to motivate people to care for this planet's ecological domains.

For Nanda and critics who share her values, religion per se is a harmful thing regardless of its form. They advocate for developing an ecological ethic that is completely secular. At issue is how realistic is it to attempt to develop a worldview regarding what I call the "ecology of the unknown"—that which lies beyond reason—by applying reason to the problem?

Nuanced Religious and Spiritual Perspectives on the Ecology of the Unknown

A central issue associated with Nanda's critique of religion is her penchant for associating the beliefs of *all* or even *most* religious people with the comparatively aggressive or, to use a Christian term, the "evangelical" beliefs and practices of religious fundamentalists. Nanda's critique of religion and ecology lacks nuance. It fails to appreciate that people may be religious and *not* orthodox, religious and *not* principally motivated to evangelize or recruit others to their faith or seek to impose it upon others, religious and *not* intolerant, religious and *not* exploitive of nature—which is to say that they may be religious *and* be tolerant *as well as* ecologically sensitive. Unfortunately, Nanda's critique reinforces the perception that many environmentalists are antireligious (Gingrich, 2007) or that there is an emerging "secular" versus "religious" orientation toward the environment (Bloch, 1998).

Pragmatically—or perhaps realistically—speaking, when religious fundamentalists and those seeking to impose a particular theological orientation upon others are sorted out from the broader constituencies of world religions, and when religious values are identified that may be expressed philosophically or in practice, then I think it is possible to conceive of religion as a source of ecological nurture as opposed to constituting the foundations for ecological crisis. Similarly, the fact that religion and theology have been and continue to be ever present in human history and culture reflects the extent to which humans will seek to find meaning beyond the bounds of rationality and science within the ecology of the unknown.

In each of the previous critiques of the world's major religions, it has been possible to identify values that require human beings to live in harmony with nature—regardless of whether these values involve ecological stewardship (as was the case for Judaism, Christianity, and Islam) or some form of nonhierarchical environmentalism based upon the spiritual unity of human beings with all of nature (as can be found in Confucianism, Hinduism, Buddhism, and Jainism). These religious faiths also promote ecological perspectives that approach cosmic proportions while espousing theological values that do not generally appear to be antithetical to simultaneously embracing a scientific cosmology such as the standard theory. Consequently, denigrating the potential for religion and theology to positively contribute toward fostering a sustainable ecological ethic among adherents is tantamount to judging all religions on the basis of the worst theology and upon the teachings and actions of the most strident religious extremists. Furthermore, it may serve to overemphasize the ecological promise of reason and science while underestimating the potential contributions of spirituality, religion, and faith.

Finally, discounting the value of religion as a force for ecological sustainability suggests a willingness to pursue ecological sustainability on purely rational-scientific grounds without securing the cooperation of the world's millions of religious people. Not only are religious ideas devalued in such a process, but even more importantly, religious people themselves are discounted and ignored. However, the most significant problem associated with ignoring or devaluing the cosmological and ecological potential of religion is the distinct likelihood that any worldwide effort to protect the planet's natural resources and life forms that excludes religion and the religious is quite simply doomed to fail. If one is truly committed to ecological action — as compared to being invested in ecological ideas — then the only pragmatic and reasonable option available is to seek solidarity for ecological sustainability from among the religions of the world.

For example, a 2004 Pew Foundation survey indicates that in the United States there is a strong consensus across all Christian denominational groups that Christianity requires its adherents to be active environmental stewards (Pew, 2004). Given this theological resolve on the part of Christians in the United States to preserve the environment, it would be foolish to pursue an ecological ethic that excluded the potential contributions of the millions of Christian faithful in the United States alone. By comparison, the pragmatic approach is to incorporate religious ecological stewardship values into a larger cooperative effort among secular and religious people of all persuasions to work together to clean up the environment and avoid further ecological degradation. This is a goal worth pursuing and an approach that requires tolerance and hope: tolerance of theological and intellectual diversity supporting ecological action and hope (or perhaps faith) that such a process will ultimately contribute to a cosmic ecology — a scientifically, spiritually, and religiously derived sense of place and purpose in the universe — that will produce more good for the creatures and environs of the world than disruption or harm.

SIX

Essential Characteristics of Nested Ecology

....

Prerequisites for System Health

....

In 2006, the University of Guelph hosted an important conference in Ontario on how to promote sustainable livelihoods for humans while ensuring ecosystem health. The meeting drew participants from around the world and its proceedings were recorded and synthesized into a final report. One of the central findings contained within this report was the assertion that "healthy ecosystems are a pre-requisite for sustainable health and well-being in human populations" (Robinson, Fuller, and Waltner-Toews, 2003, 7). In reading this statement — which appears to be an obvious truism — it occurred to me that the converse is also true. Healthy people, families, communities, and societies are a prerequisite for ecosystem/environmental health. Unfortunately, the veracity of this insight is much less apparent to many in the environmental community even though it is arguably of equal importance. Nevertheless, fully appreciating what either statement implies requires coming to an understanding of exactly what is meant by the term "health" or "healthy" as applied to communities, families, individuals, and environments.

This concern equally applies to any discussion of nested ecologies along the lines that I present in this work. While I believe it is necessary to conceive of ecology as a set of nested domains, such an approach is unlikely to produce improved ecological outcomes unless it also promotes "healthy and sustainable ecologies" at every nested level. However, since nested ecology as I articulate it is essentially a new approach to conceptualizing the place of human beings in the world, there is no existing consensus among ecologists regarding just what it means to approach each level of ecology — personal, social, environmental and cosmic — in a "healthy and sustainable" fashion. To that end, I feel the need to clarify what I have in mind when using these terms, even though doing so requires me to develop definitions in the face of a comparatively meager body of existing scholarship on these topics.

Healthy and Sustainable Communities

Following up on Robinson, Fuller, and Waltner-Toews's (2003) observation, a logical place to begin thinking about "healthy and sustainable" ecologies is at the level of "human populations," which, from my nested perspective, fall within the realm of social ecology. To date, social ecology has been primarily discussed as a community phenomenon. Unfortunately, there is a shortage of definitions available to describe healthy human communities. The crudest definition of health is the absence of illness or disease (Seedhouse, 2001, 31). By this measure, a healthy community is one in which disease is absent — not a realistic definition when considering that pathogens are also natural constituents of ecosystems. Consistent with this crude definition of health, a healthy community could also be defined as one that is not polluted or depleted by human use. Similarly, a healthy community could simply be defined as a nondegraded locale, which is similarly inadequate and unrealistic since virtually all of the planet's environs are degraded, polluted, or depleted by human activity to one degree or another (Dutu, 2004).

A more complete and functional definition for healthy communities was proffered by Leonard Duhl of the University of California at Berkeley. Duhl defines a healthy community as "one that is continually creating and improving those physical and social environments and expanding those community resources which enable people to mutually support each other in performing all the functions of life and in developing to their maximum potential" (Duhl, 2003, 94). Duhl's perspective is comprehensive yet succinct and corresponds nicely to an understanding of basic social ecology needs.

A complementary perspective on healthy communities has been proposed by the World Health Organization (WHO). WHO's vision of healthy communities is not nearly as succinct as Duhl's but is even more comprehensive, describing such communities in terms of their cleanliness and safety, long-term stability and sustainability, social supportiveness, and capacity for facilitating citizen participation and involvement. Such communities are further characterized as meeting a wide array of basic needs such as for clean water, nutrition, adequate income, accessible and affordable health care, and access to safe and meaningful work. These healthy communities also provide ready and reliable access to a wide variety of social and cultural experiences and resources. Moreover, such communities are economically diverse and vital and consistently seek to promote a sense of community connectedness and continuity.

The WHO characterization reflects an awareness of the need for sustainable ecosystems fostered by sustainable social systems. This approach lends itself to understanding what a "healthy and sustainable" community might entail. However, since it lacks brevity, I prefer to paraphrase Duhl's (2003) definition and describe healthy communities as consisting of human neighborhoods, locales, and environs that are constantly conserving and improving natural environments, continually creating and improving built and social environments, and expanding those community resources that enable people to mutually support each other in performing all the functions of life — including developing to their maximum potential.

Fortunately, it is much easier to define the term "sustainability" than it is to define what a healthy community entails. According to the Brundtland Commission (the World Commission on Environment and Development), sustainability implies meeting "the needs of the present without compromising the ability of future generations to meet their own needs" (Brundtland Commission, 1987, 43). When this straightforward definition is applied to human communities, it might be alternately defined as meeting current community needs without exhausting the capacity of future community members to meet the same basic set of needs. Upon applying this definition of sustainable communities to the previous description of what constitutes a healthy one, I propose the following composite definition of "healthy and sustainable communities":

> *Healthy and sustainable communities are constantly conserving and improving natural environments, continually creating and improving built and social environments, and expanding those community resources required to insure that current and future residents will be able to per-*

petually extend mutual support to one another and realize their maximum
potential in performing all the functions of life.

Sustainable and Healthy Families as Components
of Social Ecology

Social ecologies do not consist of communities alone. Communities are themselves conglomerates of more basic subsystems, particularly family systems. Consequently, any effort to describe a healthy and sustainable social ecology must include consideration of what a family consists of and what constitutes a healthy and sustainable family.

Families consist of a set of interdependent individuals and subsystems — including parents, siblings, extended families, etc. — and in turn interact with a plethora of external community groups and systems (von Bertalanffy, 1968). It is ironic that the bulk of what we know regarding what a healthy family looks like comes from surveying the literature on dysfunctional families and how they are to be treated. The major figures associated with this clinical research include Salvador Minuchin (Minuchin, 1998, 1986, 1974; Minuchin and Fishman, 1981; Minuchin, Nichols, and Lee, 2006), Jay Haley (2003), Virginia Satir (Satir, 1988; Baldwin and Satir, 1987), Murray Bowen (1994), and Monica McGoldrick (1998), to name but a few.

These clinicians have documented their experiences with families that were, for example, unable to provide for the basic security and nurturance needs of family members, unable to maintain clear lines of communication and authority either within the bounds of the family or along the boundary with external systems and environments, or systemically unstable and fraught with problems of physical and mental illness, sociopathic behavior, and legal difficulties. Each clinician approached these issues from a different theoretical orientation. However, there exists across all these theorists a set of common convictions that at a minimum healthy families are those that

- provide for the safety and identity needs of all family members;
- maintain clearly defined, permeable boundaries between members and subsystems in the family, as well as between the family system

and the external environment, in order to insure clear communication and to enforce lines of leadership and authority in the family;

- dynamically seek to maintain harmony and equilibrium between the external and internal demands of the family;

- promote health, personal growth, and productivity among family members;

- reliably solve personal and family problems; and

- effectively plan for and manage major life events that affect the family such as births, relocation, marriages, divorces, deaths, and illnesses.

One could claim that the very criteria that contribute to healthy families also render them sustainable. However, I would assert that there are additional criteria for insuring that family systems are sustained over time. Perhaps the most obvious is the commitment on behalf of a family's leadership and members that they want to continue to exist as a functional unit (Downs, 2004; Phelps, 2000; DeFrain, 2000; Bielby, 1992). Such an observation may seem obvious, but not every family is this purposeful or deliberate in assuring the family's prosperity and existence. Related to this characteristic is the willingness of all members of a family to build a sense of coherence and consensus among one another regarding the future of the family (Sheridan, Eagle, and Dowd, 2005; McCubbin and Patterson, 1983). Coherence and consensus go hand in hand in family systems since both qualities tend to reinforce one another such that in the absence of either quality the other could not exist.

I would further suggest that any sustainable family system must have the capacity for adapting to change (Hawley and DeHaan, 1996; Antonovsky and Sourani, 1988). All systems and subsystems constantly change over time. The family as a social system exists within a milieu of environments external and internal to its boundaries. Consequently, when the environment (external or internal) changes, the dynamics of the family shift, resulting in subsequent changes among family subsystems and along with these changes, changes among all of the family boundaries (external and internal).

While the family-system literature provides a clear assessment of what healthy families look like and suggests characteristics that contribute to family sustainability, it fails to provide a succinct definition of what a "healthy and sustainable family" might entail. Since families (as will be discussed shortly) are the foundational system for communities and, by extension, social ecologies, I would like to present a tentative description of a "healthy and sustainable" family system that might fruitfully be applied later in the text. The definition I propose is reflective of the bulk of the clinical literature that characterizes a healthy and sustainable family as

> *a complex kinship system organized into subsystems around a recognized leader or leaders (typically parents) that enforces family role and authority boundaries; promotes family coherence, identity and health; insures the ongoing success and longevity of the family as a unit; successfully satisfies the basic needs and interest of its members, resolves family problems and disputes; and adapts to changes within and beyond the family environment.*

Healthy and Sustainable Personal Ecologies: Lifestyles

This twofold perspective regarding healthy and sustainable communities and families is most useful when applied to social ecology. However, there is also a need to develop an understanding of "healthy and sustainable" applicable to personal ecology. To that end, I propose exploring personal ecology from the perspective of individual lifestyles and suggest what a "healthy and sustainable lifestyle" might look like.

One way to approach this issue is to begin with those characteristics that are believed to be indicative of an unhealthy lifestyle, and on the basis of this categorization, posit an opposite set of healthy characteristics. Researchers from Norway's University of Tromsø (Olsen, Richardson, and Menzel, 1998) have suggested an intriguing way to think about unhealthy lifestyles within families, proposing three sets of contributing risk factors, to include "those that refer to a person's *relations to other people* in society (e.g., having children); those that place the person *in a causal relationship with the illness,* i.e., the extent to which a particular illness might have been influenced by their own actions (e.g., smoking), and those 'embodied' in a *person's self* physically, intellectually or attitudinally (e.g., gender)" (Olsen, Richardson, and Menzel, 1998, 3).

In the United States, the principal behaviors that produce the most ill health and disability include tobacco use, sedentary lifestyle, obesity, and unhealthy diet (USDHHS, 2007; Reeves and Rafferty, 2005). These behaviors occur across all three of the dimensions proffered by Olsen, Richardson, and Mendel, suggesting that healthy lifestyles include making adjustments to one's attitudes and ideas regarding health, making changes in one's habits and personal behaviors — particularly those such as smoking that directly produce deleterious health effects — and changing the way that one interacts socially that may be unhealthy. Such a perspective transforms the issue of healthy lifestyles into one that is reflective of basic choices rather than primarily genetically or environmentally determined.

In contrast, a sustainable lifestyle, from a purely human health perspective, involves the capacity of a person to consistently engage in healthful behaviors. Obvious factors in this regard include the person's economic resources, education, and the larger set of social systems in which they function — systems that may or may not contribute to their capacity to sustain a healthy lifestyle. Likewise, they may have genetic dispositions to disease and disability, and these too may serve to make it more difficult for them to sustain a healthy lifestyle. Given the influence of these factors upon human health, realizing a sustainable lifestyle may additionally require that a person have reasonable access to the requisite social services, health care, and informal social support resources.

The capacity to sustainably engage in healthy behaviors is also a function of environmental factors. Obviously, polluted and depleted environments detract from a person's ability to pursue a healthy lifestyle. However, humans are not always victims of their environments. In reality, many people possess the capacity to protect and improve upon natural environments, and in so doing, improve their own health and well-being. If one assumes — based upon Olsen, Richardson, and Mendel's (1998) attitudinal choice model — that a person has the wherewithal to think, act, and interact in a healthy fashion with other people and also has the requisite resources to do so, then they likewise have the capacity to do so in regard to other ecological domains, particularly in regard to natural ecosystems and environs. Since natural environments play an important role as determinants of whether people can realize and sustain healthy lifestyles, they must be included in any consideration of human lifestyles and their ecological sustainability.

Arriving at a functional definition of what constitutes this kind of environmentally and ecologically sustainable lifestyle is relatively straight-forward. The Economic Commission for Europe (ECE) of the United Nations defines a sustainable lifestyle as "a way of life which does not contradict the aims of preserving natural resources of present and future generations as well as health. Ecological requirements of 'sustainable life-style' essentially consist in protecting resources and safeguarding ecologi-cal balance against irreversible damage. Alongside these ecological re-quirements, societal ones must be taken into account as well: social and personal stability — health and well being of every individual" (Lang-schwert, 1998, 2). Moreover, "a healthy lifestyle is about striving to obtain a reasonable balance between enhancing one's personal health, the health and well-being of others, and the health of the community" (Lyons and Langille, 2000, 14).

Having arrived at a basic understanding of what a "healthy lifestyle" might look like as well as what a "sustainable lifestyle" entails, we can conflate the content of these two definitions to arrive at a definition of a "sustainable and healthy lifestyle" as

> *a way of life that does not contradict the aims of preserving natural re-sources of present and future generations and that seeks a reasonable bal-ance between the enhancement of personal health and satisfaction, the health and well-being of others, and that of the broader community.*

Subsequently, when I refer to "healthy and sustainable" personal ecolo-gies, I will do so with this definition in mind.

Healthy and Sustainable Environmental Ecologies

Having clarified what I mean by "healthy and sustainable" as it applies to social and personal ecologies, I now turn to the dimension of environmen-tal ecology and define what it means to have a "healthy and sustainable environment." Once again, a clear definition of what constitutes a healthy and sustainable environment is lacking in the scientific and professional literature. However, definitions are to be found for "healthy environ-ments" and "sustainable environments." Arriving at a suitable definition regarding a sustainable and healthy environment involves reviewing the best available definitions for both terms and integrating their content into a composite definition.

Healthy environments may be defined in terms of ecosystems capable of maintaining productive and sustainable populations of indigenous species (Rapport et al., 2002), as "ecological systems capable of maintaining organization, autonomy, and resistance to stress" (Costanza et al., 1992) or as an ecosystem "that persists, maintains vigor (productivity), organization (biodiversity and predictability), and resilience (time to recovery)" (Costanza and Mageau, 2000). An alternative approach is to explore definitions of "environmental health" as compared to "healthy environments." For instance, environmental health can be defined as "protection against environmental factors that may adversely impact human health or the ecological balances essential to long-term human health and environmental quality, whether in the natural or man-made environment" (NEHA, 1996). Comparatively, the World Health Organization (WHO) defines environmental health as consisting of "those aspects of human health, including quality of life, that are determined by physical, chemical, biological, social and psychological factors in the environment" (USDHHS, 1998).

A set of divergent definitions of environmental health emanating from the National Center for Environmental Health of the U.S. Centers for Disease Control (CDC), the Institute of Medicine (IOM), and researchers at the University of New Mexico are also applicable. The CDC defines environmental health as "the discipline that focuses on the interrelationships between people and their environment, promotes human health and well-being, and fosters a safe and healthful environment" (Robinson, 1992, E-15). Alternatively, the Institute of Medicine (IOM) defines environmental health in terms of "freedom from illness or injury related to exposure to toxic agents and other environmental conditions that are potentially detrimental to human health" (Pope, Snyder, and Mood, 1995). Meanwhile researcher Larry Gordon at the University of New Mexico construes environmental health as linking "the environmental quality of both the natural and built environments, with the level of public health and well being" (Gordon, 1998).

As illustrated, the available research literature provides many more definitions of environmental health than it does of healthy environments. The problem, of course, is that definitions of environmental health tend to emphasize human needs and health, whereas definitions of healthy environments focus upon the health of the environment per se. In this instance, the definitions of healthy environments are most applicable to an understanding of environmental ecologies. Consequently I favor the perspective of Costanza and Megeau (2000), who define healthy environments as ecosystems that persist, maintain *vigor* (productivity), *organization* (biodiversity and predictability), and *resilience* (time to recovery).

By comparison, sustainable environments consist of "ecosystem components and ecological processes that enable the land, water, and air to sustain life, be productive, and adapt to change" (Hammond et al., 1996, 2–18) or in the form of complete ecosystems "that will remain healthy and thriving for the long term" (Robbins, 1996). In this case, some of the same characteristics that render an ecosystem healthy also appear to render it sustainable. When these divergent definitions are conflated, a healthy and sustainable environment might well be defined as

> *an ecosystem that persists over time; remains productive; exhibits biodiversity, resilience, and adaptivity to change; and that can be expected to continue doing so into the foreseeable future.*

Healthy and Sustainable Cosmic/Spiritual Ecologies

The final application of the terms "healthy and sustainable" is reserved for cosmic ecology. This is a more challenging application in that cosmic ecology encompasses scientific, philosophical, and spiritual dimensions. Moreover, unlike environmental ecology, which operates at the level of ecosystems, cosmic ecology involves the entire biosphere, and unfortunately there is little in the way of any clear consensus regarding what constitutes a healthy and sustainable biosphere. Perhaps the nearest thing to an assertion of planetary sustainability comes from the work of a single author.

David Wilkinson, a proponent of Gaia and planetary ecology, in *Fundamental Processes in Ecology* (2006) identifies seven fundamental processes he believes are basic for sustaining life on the planet, including:

1 Energy flow: energy consumption and waste product excretion

2 Multiple guilds: multiple aggregates of organisms that share a similar form of life, such as *autotrophs* (organisms producing complex

organic compounds), decomposers (organisms consuming dead organisms and matter), and parasites

3 Tradeoffs: specialization versus generalization, leading to biodiversity within guilds

4 Ecological hypercycles: an integrated cyclic reaction network — cycles within cycles

5 Merging of organismal and ecological physiology: as life spreads over the planet, biotic and abiotic processes interact so strongly as to be inseparable

6 Photosynthesis: a likely though not inevitable occurrence in most biospheres

7 Carbon sequestration: redistribution of carbon from the air to the soil

Wilkinson claims that life on Earth would be impossible without these seven fundamental processes; therefore, any sensible human approach to cosmic ecology must seek to insure that they continue indefinitely.

Wilkinson's work is cutting-edge in the ecological sciences, and his assertions await scientific confirmation by other researchers — this despite the fact that they are very much grounded in what we currently know about how ecosystems interact with one another across the planet. In fact, a growing number of scientists have initiated ambitious research efforts in the interest of discovering exactly how the biosphere as a living entity functions (Copley and Wessman, 2007; Bowman, 1998). While awaiting these efforts to come to fruition, however, I would suggest that we embrace Wilkinson's seven fundamentals as standards for biospheric sustainability and define a stable and sustainable planet Earth as

a complex network of interacting ecosystems combining to create a biosphere integrated by a set of fundamental ecological processes without which life on the planet would cease to exist.

Consequently, insuring the stability and sustainability of the biosphere requires that all of the fundamental processes for life on the planet be monitored and vouchsafed in the interest of sustaining life as we know it on Earth. Unfortunately, this is the best that can be done at the present time in terms of defining what a sustainable cosmic ecology entails. We will have to await further research findings to make an assertion regarding what constitutes a healthy biosphere.

In making such assertion, I recognize that the biosphere extends upward to the outer limits of the planet's atmosphere. What makes any consideration of ecological sustainability particularly difficult, however, is the fact that while humans currently exert an ecological influence upon the upper atmosphere and in the nearer reaches of outer space, we lack the capacity to accurately measure the impact of our activities. Given the inherent difficulties in making such an environmental needs assessment, it is not possible to provide a clear definition of "healthy and sustainable," "outer space," or "upper atmosphere" at this time.

On the other hand, it is possible to address this issue from the perspective of the spiritual cosmology implicit within cosmic ecology as "the ecology of the unknown." Comparatively little has been written that suggests what may or may not characterize healthy spirituality or religiosity in the face of the cosmic unknown. Perhaps the most popular book on this topic is Melanie Svoboda's *Traits of a Healthy Spirituality* (2005) in which she lists and discusses twenty traits of healthy spirituality ranging from self-esteem, wonder, and friendship to interdependence, perseverance, freedom, and generativity (creating life). Svoboda's twenty traits are presented from a distinctly Roman Catholic perspective, so some traits may or may not prove to be useful when applied ecologically. Nevertheless, a number of those she discusses — such as wonder, tolerance, interdependence, perseverance, generativity, balance, gratitude, and commitment — are consistent with characteristics cited by other less ecclesiastically oriented writers. For instance, F. A. Prezioso, writing about the usefulness of spirituality observes:

> Spirituality is a quality that belongs exclusively to the human animal. . . . It's the life energy, the restlessness, that calls us beyond "self" to concern for, and relationships with, others and to a relationship with the mysterious "other." Spirituality is our ability to stand outside of ourselves and consider the meaning of our actions, the complexity of our motives and the impact we have on the world around us. It is our capacity to experience passion for a cause, compassion for others and forgiveness of self. Spirituality is a process of becoming, not an achievement; a potential rather than a possession. (Prezioso, 1987, 233)

Prezioso's assertion is remarkably useful from an ecological perspective — a truly unusual outcome when one considered that it was written for a wholly different purpose than articulating an ecological ethic. His article is particularly applicable to the extent it specifically identifies healthy characteristics of spirituality such as extending concern and establishing relationships with others, establishing a relationship with the "mysterious other" (the divine), achieving empathy by standing "outsides of our-selves," appreciating the complexity of the world, coming to appreciate human impact upon the world, experiencing passion for a cause larger than ourselves, demonstrating compassion for others, acquiring forgive-ness of self and others, perceiving oneself and the world as "becoming" not complete, and relating to self, others, and the world as a "potential" rather than a possession.

Yet another useful statement pertaining to healthy spirituality ema-nates from a most unusual source — the renowned civil-rights leader and pastor Dr. Martin Luther King Jr. (1958). In an interview with *Ebony* magazine, Dr. King made the following observation regarding the value of religious faith: "One of the things that a positive and healthy religious faith gives an individual is a sense of inner equilibrium which removes all basic worries. Religion does not say an individual will never confront a problem, or that he will never worry about anything; it simply says that if the individual is sufficiently committed to the way of religion, he will have something within that will cause him to transcend every worry situation with power and faith" (King Jr., 1958, 445). King's comments add addi-tional characteristics to what constitutes healthy religion and spirituality, including acquiring a sense of inner equilibrium, alleviating rumination and worry, and contributing to a sense of equanimity from being able to transcend the constraints of the present in order to perceive the world in a transcendent and eternal fashion.

Spirituality can also be regarded as psychologically healthy. Clark F. Vaughn, writing from a transpersonal psychology perspective (the psy-chological discipline that studies the transpersonal, transcendent, and spiritual attributes of the human experience), presents an extensive set of characteristics that he believes are associated with healthy spirituality, including authenticity, letting go of the past, facing fears, acquiring insight and forgiveness, love and compassion, experiencing community, inner awareness and peace, and liberation from fear and ignorance (Vaughn, 1991, 117). As extensive as Vaughn's criteria are, I would hasten to add three additional characteristics that I believe to be particularly useful in defining a healthy spiritual ecology:

- Otherness: generating an awareness of all life as wholly other and divine, thereby engendering regard for all life and compassion for the needs of all creatures

- Transcendence: motivating people to recognize the limitations of reason and science and to consider that there may be realities that transcend the known

- The common good: enabling people to respect the common good as it relates to all people and creatures

This array of ideas regarding what constitutes healthy spirituality is most useful as a reservoir from which an ecologically oriented definition of spirituality can be drawn. To that end, I would like to propose the following ecologically oriented definition of healthy spirituality:

> *Healthy spirituality is a life perspective that engenders inner peace and patience, acknowledges the existence and legitimacy of other persons, species, and communities — social and natural — and conceives of these as reflecting the presence of a creative and ever-present Other reflected in all that is to be found on the planet and throughout the cosmos. Such an orientation commands respect for all creatures and communities; demands caring, love, and succor for all people and creatures; engenders humility regarding one's place in the world; and reorients the human relationship to the world from one of "possession" to "participation" and "potential."*

Sustainable spirituality, by comparison, is defined by neo-humanist Roar Bjonnes as "the idea that true progress is movement toward inner fulfillment, toward self-realization" (Bjonnes, 2003). Comparatively, Mark Wallace of Swarthmore College takes a very different approach, observing: "Sustainable spirituality is an exercise in rhetorical reason rather than a scientific enterprise in the narrow sense of the term. Its goal is to motivate all persons to live responsibly on the earth . . . The point of sustainable spirituality is to imagine the world as a communitarian family of beings that mutually depend upon one another in order to liberate sisterly feelings for the many life forms that populate the earth" (Wallace, 2003, 602). In a later piece, Wallace further elaborates upon this concept, asserting that sustainable spirituality "is an earthen spiritual vision concerning the deep interrelationships of all life-forms on the planet and the concomitant ethical ideal of preserving the integrity of these relationships through one's religious and political practice. Sustainable spirituality offers its practitioners a powerfully useful root metaphor — the image of all life as organically interconnected — that can enable a fresh reappraisal of the debate between advocates for environmental justice and biocentric conservationists" (Wallace, 2005).

Wallace's ideas are derived from those of Charlene Spretnak, who describes sustainable religion as pluralistic, post-humanist, postmodern, and post-patriarchal — comparing this contemporary religious orientation to a form of spirituality that entails "the focusing of human awareness on the subtle aspects of existence, a practice that . . . reveals to us profound interconnectedness" (Spretnak, 1986, 24). Consequently, the central tenet associated with sustainable spirituality as articulated by Spretnak and Wallace is that of interconnectedness.

Building upon Wallace and Spretnak's vision of an interconnected spirituality and my own and Vaughn's perspective regarding healthy spirituality, I propose the following definition of healthy and sustainable spirituality:

> *Healthy and sustainable spirituality is a life perspective that embraces the interconnectedness of all that is known and unknown in the world. It engenders inner peace and patience, acknowledges the existence, legitimacy, and interrelatedness of other persons, species, and communities — social and natural — and conceives of these as reflecting the presence of a creative and ever-present Other reflected and infused in all that is to be found on the planet and throughout the cosmos. Such an orientation demands respect for all creatures and communities; caring, mutuality, love, and succor for all people, creatures, and communities; engenders humility regarding one's place in the world; and reorients the human relationship to the world from one of "possession" to "participation," "potential," and "kinship."*

From Healthy and Sustainable to Pragmatic

I believe that a healthy and sustainable nested ecology maximizes the likelihood that ecological values will ultimately be reflected in human behavior. This should especially be true if health and sustainability are ecologically experienced personally, socially, environmentally, and spiritually. Even so, if ecological values are to truly be reflected in behavior, then a nested approach to ecologically relating to the world must also be perceived as thoroughly pragmatic — otherwise it will share the fate of any of a number of other ecological philosophies that add to the content of philosophy textbooks but make little or no practical contribution to ecological improvement.

Pragmatism possesses a number of advantages that are most useful when applied to a nested ecological perspective. For instance, Joseph Margolis (2003) has lauded pragmatism's disdain for philosophies that would uniformly impose the methods of the physical sciences upon every field of inquiry. In so doing he characterizes pragmatism as an orientation "freed from every form of cognitive, rational, and practical privilege, opposed to imagined necessities . . . , committed to the continuities of animal nature and human culture, confined to the existential and historical contingencies of the human condition, and open in principle to plural, partial, 'perspectived,' provisional, even non-converging ways of understanding what may be judged valid in any and every sort of factual and normative regard." Pragmatism is further described as "a social construction of the world determined by the sense people give to concepts and their sensations, expectations and beliefs about the value of both knowledge and the inquiry process" (Durand and Vaara, 2006).

The fact that pragmatism is grounded in human experience is its principal advantage. Pragmatic models can be adjusted as needed to accommodate changed experience and are also responsive to the scientific method — although not inextricably bound to it or to any other particular theory or ideology that is not reinforced by experience. This allows pragmatism to adapt and to change as circumstances and experience dictates.

The nested approach to ecology articulated in this work is thoroughly grounded in human experience as reflected in scientific observations. It assumes a human perspective on the world, including the wide variety of ways in which humans interact with one another and the world around them. Pragmatism concerns itself with how human beings function biologically, psychologically, sociologically, anthropologically, scientifically, economically, and politically. It assumes a necessarily human perspective on how people relate to the variety of ecological domains found throughout and within the world. Finally, it assumes that human beings are a legitimate part of this world, as is human culture and society, and articulates a unified and integrated ecological perspective that rightfully places human beings and their social systems *within, not above or beyond,* natural ecological systems. Located within nature, and appropriately so, humans are empowered in this model to live ecologically in the fullest sense of the word.

Nested Ecology as a Pragmatic Worldview

Nested ecology serves as a practical and pragmatic worldview (cosmology) that allows ordinary people to think of themselves, their lives, their family and friends, their community, nation, and the world as a set of interrelated and interdependent households — such that what goes on in any one of these myriad households affects all others. It is also a worldview that revolves around an obvious but often overlooked assumption — namely, that individual human beings perceive nature from the confines of their own subjectively experienced lives, and that if their lives are amiss or under stress their capacity to recognize or appreciate the reality and demands of nature is dramatically impaired.

While this assumption may be patently obvious to those with backgrounds in the psychological or social sciences, it is an assumption all too easily overlooked or underappreciated within the basic sciences and among ecologists, environmentalists, and activists. People simply cannot appreciate the needs of nature or their relationship to natural ecosystems if they are hungry, ill, unhappy, stressed, or exposed to violence or if they function within a social environment that is devoid of personal meaning or value. Likewise, human beings are not likely to be cognizant of their larger ecological responsibilities and relationships if they are distracted in their lives — regardless of whether that distraction emanates from the pursuit of basic survival, earning a living, securing a better community, or anything else.

Some may discount the importance of these insights as no more than the problem of anthropocentrism restated. However, anthropocentrism is not simply an attribute that can somehow be eliminated from human behavior or community. Anthropocentrism is also the process whereby human beings care for their personal needs, address the needs of others, and adapt to the constraints of the world around them. It is the human version of what all other creatures do in pursuit of survival and species perpetuation. Ultimately anthropocentrism is not an optional or aberrant trait. It is in fact a basic survival instinct that is played out personally, socially, and environmentally. In recognition of that basic fact I have chosen to present a nested-ecology worldview that begins with the world-view of an ordinary person — everyman — and moves ever outward until it ultimately encompasses the ecology of the cosmos and the unknown.

I seek to present a cosmology or worldview that is pragmatic in that it approaches human ecological philosophy and ethics from the perspective of the individual person in hopes that such an orientation might be useful in helping people acquire a truly ecological perspective regarding themselves and all else in the world, as well as in the interest of suggesting steps that might be fruitfully undertaken to expand the scope of ecological thought and action in the interest of simultaneously improving the lot of humankind and those of natural environs and creatures of this planet and perhaps beyond.

I realize, however, that while there is a pressing need for a practical and pragmatically conceived ecological ethic, the process of successfully articulating an ethic that will actually be utilized by people is a daunting task. Andrew Light and Eric Katz (1996) took on this very issue more than a decade ago and assembled a broad array of ecological thinkers in the interest of articulating a pragmatic ecological ethic. They did so recognizing that prior to their effort, ecological ethicists and philosophers had exerted scarcely any influence on policymakers at all. Unfortunately, it would seem that since the publication of Light and Katz's book little has changed.

In writing this book my goal is not to directly or immediately influence environmental policies. Instead I am interested in influencing individuals' ideas about themselves and the ecological realms they reside in, hoping that by incrementally changing their worldviews I can begin to change the way they approach their lifestyles and interactions with other people, creatures, and ecological systems. Pursuing such a goal makes me a pragmatist in the great American tradition of pragmatism, because I develop my ecological philosophy and ethics on the basis of how people actually relate to themselves, others, and the world. My approach is grounded in the assumption that humans have evolutionarily adapted — as has every other creature — to the world they live in and simply *must* define themselves and their species-specific culture in relationship to other people, creatures, and environments.

Philosopher Kelly Parker summed up the assumptions of American pragmatism quite well when he observed: "The founders of pragmatism recognized the philosophical implications of evolutionary theory. The characteristics and activities of any organism are always understood in light of the organisms' relations to its environments. The human capacities of thinking and knowing are no exception. Consciousness, reason, imagination, language, and sign use (*mind,* in short) are seen as natural adaptations that help the human organism to get along in the world" (Parker, 1996, 23). My nested ecological worldview is grounded in an effort to articulate an ethical and philosophical orientation to the world that is consistent with what we have observed regarding an "organism's relations to its environments" (Parker, 1996, 23).

However, the pragmatism reflected in my nested ecological approach is not simply the product of my emphasis upon the interrelatedness of all creatures and environments. It is also grounded in what some might call my anthropocentric insistence that humans are evolutionarily and biologically compelled to view the world and formulate their relationship to it within the limits of what their minds are capable of perceiving. Once again, to quote Parker, "To a pragmatist, the concept of a world, entity, or property existing apart from the ordering influence of the mind is strictly *meaningless*" (Parker, 1996, 24). While this might be a self-evident observation to many, it reflects the extent to which it is virtually impossible and impractical to articulate any philosophical or ethical value that is not principally anthropocentric in its orientation. Consequently, I endeavor to articulate an ecological philosophy, cosmology, and ethic that is unabashedly anthropocentric in the interest of influencing human beings in the direction of becoming much more accommodating and complementary in their interactions with the other "households" and "communities" they share the planet with.

An additional insight emanates from this pragmatic ecological worldview. If we only know the world through our interactions with other people, creatures, and environments, and if the reality of these interactions is experienced via the neural pathways of our minds, then we are no different than any other sentient creature and are as constrained as they are to interact with the world and create a species-specific culture on the basis of these interactions and our cognitive perceptions of them. What this means is that we are as natural as any other species when we interact with the world on the basis of our mental perceptions. Every other sentient creature does likewise. Similarly, when we create our societies and cultural structures and artifacts, these too are completely natural and ecological—as natural as termite mounds are to termites, nests to birds, honeycombs to bees, and dams to beavers.

To assume a pragmatic approach to ecology, as I do with my discussion of hierarchically nested ecological systems, is to approach humans as ecological beings residing within the natural world and *legitimately* interacting with and shaping it as does every other living creature. The upshot of this philosophy is that human beings do not need to somehow apologize for being human within the natural world. We only need to apologize when we knowingly and unnecessarily impose the needs of our species' households over the legitimate needs of the households of other species.

The Problem of Hierarchy

There are obstacles to be overcome in asserting a pragmatically oriented nested ecological worldview. Chief among those obstacles is the issue of hierarchy. Nested ecology is a hierarchical model, though one that is more akin to the hierarchies found among natural environs than similar to those found within human society. Moreover, nested ecology is conceptually very much in keeping with the philosophical values of social ecology. However, unlike those social ecologies espoused by Murray Bookchin and his intellectual disciples, it does not categorically eschew the value of hierarchy as an ecological construct or value. In this way, nested ecology as an approach to social ecology is at odds with many of the essential values of social ecology's creator, Bookchin.

At the heart of Bookchin's philosophy is his belief that any effort to save the planet must begin with improving the lot of humanity and protecting nature from the excesses of human culture. I share these sentiments. However, I disagree with Bookchin regarding the underlying factors that contribute to and sustain human suffering and oppression and environmental degradation. Bookchin approaches social ecology from the philosophical perspective of an anarchist—in the tradition of Peter Kropotkin (1902)—and as a dedicated communist steeped in the thought of Marx and Engel (Bookchin, 2003). As such, he perceives the ills and excesses of humanity as emanating from hierarchy, which he associates with the domination of an elitist capitalist bourgeois upper class over a comparatively impoverished and dependent proletariat class. Bookchin asserts that the domination of the proletariat by the bourgeois capitalists and a similar bourgeois domination over the environment is realized through the enforcement of a top-down social hierarchy system along the lines of that described by Max Weber (1947).

Social ecology, as conceived by Bookchin, is required to abolish hierarchy. The method by which Bookchin would achieve this end is not only informative of his unique vision of social ecology but also provides an excellent departure point for understanding nested ecology as a contemporary social ecology approach. According to Bookchin:

> Hierarchy had to be abolished by *institutional* changes that were no less profound and far reaching than those needed to abolish classes. This placed "ecology" on an entirely new level of inquiry and praxis, bringing it far above a solicitous, often romantic and mystical engagement with an undefined "nature" and a love-affair with "wildlife." Social ecology was concerned with the most intimate relations between human beings and the organic world around them. Social ecology, in effect, gave ecology a sharp revolutionary and political edge. In other words, we were obliged to seek changes not only in the objective realm of economic relations but also in the subjective realm of cultural, ethical, aesthetic, personal, and psychological areas of inquiry. (Bookchin, 2003, 7)

Like Bookchin, my nested-ecology approach is "concerned with the most intimate relations between human beings and the organic world around them." I likewise seek "changes not only in the objective realm of economic relations but also in the subjective realm of cultural, ethical, aesthetic, personal, and psychological areas of inquiry." However, I do so without embracing Bookchin's disdain for hierarchy because, unlike him, I recognize a variety of hierarchical forms in nature, not all of which are toxic and dysfunctional. In essence, I propose a nested-ecology approach that is socially ecological in spirit, hierarchical in structure, but "naturally" rather than "socially" or "formally" hierarchical.

Hierarchy in Nature

Biologists have perennially commented upon the hierarchical organization of nature. In fact, hierarchy as been associated with the scientific study of nature since at least the eighteenth century with the publication of *Systema Natura* by Carolus Linnaeus (1735) and later in the nineteenth century with the publication of Charles Darwin's *On the Origin of Species* (1882). In the modern era, an emphasis upon the hierarchical order of nature can be found in the work of biologist Paul A. Weiss, who describes it as a "demonstrable descriptive fact" (Weiss, 1969, 4), and Ludwig von Bertalanffy, who asserts that hierarchical structure "is characteristic of reality as a whole" (von Bertalanffy, 1968, 87).

Weiss and von Bertalanffy are not alone in their assertions regarding hierarchy. Their views have been echoed by a number of prominent biologists, scientists, and logisticians to include James Feibleman (1954), Michael Polanyi (1968), Howard Pattee (1973), John Lionel Jolley (1973), and Donald Campbell (1974). However, the most thorough and influential analysis of hierarchy was provided by Herbert Simon, a nonbiologist who actually won a Nobel Prize in Economics in 1978 for his work on hierarchy and decision making in organizations. Simon's exposé on hierarchy appeared in his now famous book *The Sciences of the Artificial* (1969), and some his thoughts in this regard are particularly applicable to an understanding of nested ecology as a functional approach to hierarchy.

Simon defines hierarchy as "a system that is composed of interrelated subsystems, each of the latter being in turn hierarchic in structure until we reach some lowest level of elementary subsystem" (Simon, 1969, 184). Simon contrasts this understanding of hierarchy with a "narrower" definition popularly applied within bureaucratic organizations—a form that has also been characterized as social hierarchy (Bull, 1977; Schein, 1975; Fromm, 1956). He describes such hierarchy as automatically entailing subordination. More specifically he observes that, "the term [hierarchy] has generally been used to refer to a complex system in which each of the subsystems is subordinated by an authority relation to the system it belongs to. More exactly, in a hierarchic formal organization each system consists of a 'boss' and a set of subordinate subsystems. Each of the subsystems has a 'boss' who is the immediate subordinate of the boss of the system" (Simon, 1969, 185).

Simon rejects this popular understanding of hierarchy as inadequate to describe complex systems and instead proposes a model that does not involve subordination whatsoever. In fact, he goes on to observe that "in human organizations the formal hierarchy [involving outright subordination] exists only on paper; the real flesh-and-blood organization has many interpart relations other than the lines of formal authority." Ultimately, Simon substitutes the common subordination model of hierarchy with a non-subordination definition in which hierarchy is simply conceived of as "all complex systems analyzable into successive sets of subsystems," while describing subordinated hierarchy as "formal hierarchy" (Simon, 1969, 185).

When I considered writing about nested ecology, I had Simon's definition regarding complex systems in mind — particularly the illustration he used to describe hierarchy, which entailed a "set of Chinese boxes" illustrative of hierarchy in complex systems:

> In application to the architecture of complex systems, "hierarchy" simply means a set of Chinese boxes of a particular kind. A set of Chinese boxes usually consists of a box enclosing a second box, which, in turn, encloses a third — the recursion continuing as long as the patience of the craftsman holds out. The Chinese boxes called "hierarchies" are a variant of that pattern. Opening any given box in a hierarchy discloses not just one new box within, but a whole small set of boxes; and opening any one of these component boxes discloses a new set in turn. (Simon, 1973, 4–5)

It is this image of a set of nested boxes that led me to conceive of ecology as a set of nested "households" — each situated within the other and each and every household interdependent and interrelated with one another. In conceptualizing such a set of nested ecologies, I, like Simon, conceived of a hierarchical relationship between the various ecological levels but did not consider these hierarchically organized subsystems as in any way being in a subordinate relationship to one another.

I suspect that when Bookchin and other ecologists and environmentalists decry hierarchy, what they are in fact protesting are hierarchical systems and relationships involving subordination of one group, one interest, or one ecological system to the authority and power of another system, class, or interest. However, as Simon illustrates with his observation about hierarchy in "real flesh-and-blood organizations," it does not follow that all hierarchy is inevitably subordinated in the fashion of "formal" or popularly conceived "social" hierarchy. Rather than characterizing nested ecology as hierarchically subordinate, I would prefer to use a term that Bookchin himself proffers and describe nested ecology as exhibiting hierarchical "complementarity" (Bookchin, 1986, 75).

Nested Hierarchies and Nested Systems

The concept of nested hierarchies is certainly not new and can be found in the works of Linnaeus (1735) and Darwin (Wilson, 1997; Darwin, 1882). Since its initial introduction, nested hierarchy has been discussed by any number of contemporary scientists, including theoretical ecologists Robert V. O'Neill, Donald L. DeAngelis, and Jack B. Waide and botanist Timothy Allen (Allen 1996; O'Neill, DeAngelis, Waide, and Allen, 1986), geographer Robert G. Bailey (2002), and landscape architects Frans Klijn and Helias Udo de Haes (1994), to name but a few. It has also been applied to ecosystems since they are in fact "open systems" that exchange energy with surrounding systems and consist of a hierarchy of systems nested within one another (Allen and Starr, 1982). Moreover, in recent years, the concept of nested hierarchies has generated some controversy, thanks to biochemist Douglas Theobald (2004) of Brandeis University, who asserts that the very presence of nested hierarchies in nature suggests that all living things originated from a single living species.

The most sophisticated discussion of nested hierarchy in an ecological sense, however, is that presented by eco-philosopher Bryan G. Norton. Norton offers a theory of what he calls "adaptive ecosystem management" that is in part built around a nested hierarchy of ecological subsystems that includes human beings and human culture. This is an orientation remarkably similar to that which I propose in nested ecology — although Norton's was developed for a much different purpose and at a level of analysis that largely falls beyond the scope of this book. Norton delves into the underlying meaning of Aldo Leopold's call to "think like a mountain" in defense of the environment and interprets this to involve thinking across time and scale. Leopold's admonition came from his professional perspective as a natural resource / game manager. Norton describes Leopold's approach as managerial in scope, requiring him to put himself in the place and perspective of the mountain upon which all ecological activity occurs. From this vantage point, "thinking like a mountain" requires one to locate oneself "on the mountain," observing the mountain from within its context, a context that is constrained on all sides by "dynamics unfolding on multiple scales" (Norton, 2005, 222). More specifically, Norton recognized that to think as a mountain entails recognizing "the importance of multiple temporal scales and the associated hidden dynamics that drive them" as well as realizing that "normally slow-scale ecological dynamics, if accelerated by violent and pervasive changes to the landscape, can create havoc with established evolutionary opportunities and constraints and threaten the society with extinction" (Norton, 2005, 222).

Norton's insight into Leopold's allegory of "thinking like a mountain" was that ecosystems consist of a set of nested subsystems that often vary in terms of their scale and temporal change dynamics. These nested hierarchies consist of "layered subsystems and supersystems, with the smallest subsystems being the fastest-changing, and the larger systems changing more slowly, providing environments for the subsystems" (Norton, 2006). The adaptive cycles of these ecological systems have been described by Gunderson as hierarchically nested and spatially/temporally dynamic (Gunderson, Holling, and Light, 1995). Instability within a set of these nested and dynamically related ecological subsystems can be readily introduced by human activity, especially when ecological changes are produced at rates faster than those that can be absorbed by larger, more slowly adapting subsystems.

Norton's nested, hierarchical, and scalar conceptual approach to ecological systems is consistent with Gunderson and Holling's (2002) concept of "panarchy," which is concerned with the impact and interaction of hierarchically lower, smaller, faster-changing scale levels, as well as the comparably larger, slower supra-regional systems. Panarchy focuses upon changes and rates of change among nested cycles and subsystems within larger macro-ecosystems and serves as an integrative theory concerned with the etiology and course of change within adaptive ecological systems

(Kinzig et al., 2007). These changes consist of ecological, biological, social, and economic variables that may produce local, regional, or global impacts, which in turn may produce temporary, sustained, or even evolutionary effects.

Nested Ecology for Everyman and Everywoman

Norton's work is important not only because it translates Leopold's simile into an ecological management and assessment tool but also because it pragmatically reflects contemporary knowledge and expertise regarding the interrelationships of nested ecological subsystems that operate upon divergent scales of change over time. However, as useful as these tools are to guiding humans in their interactions with complex ecosystems, they serve a purpose beyond the scope and focus of my own work. I am concerned with humans in the form of individual persons who will make individual choices and decisions. I believe that it is necessary to reassert human beings into environmentalism and acknowledge that they are an important part of the world's nested ecological systems and integral to protecting the world's ecosystems and life forms from unnecessary destruction (Grizzle, 1994).

So conceived, my principle concern is how to instill an ecological sensibility in such individuals that will serve to motivate them to regard nature and natural environments as an integral part of *their own* households. Ultimately, what I am interested in realizing through this nested ecological approach is the integration of everyman and everywoman into a set of nested ecological systems such that they would come to consider their own personal household as vitally interrelated and integrated into the households of others within the local community, the state and nation, the world community, and the broader system of natural environments and ecosystems that define and sustain all living communities and beings.

My ecological approach begins at the psychological level, proceeds to consider society as a political, sociological, and economic phenomenon, moves on to embrace the physical, chemical, biological, and ecological sciences, and concludes in the realms of philosophy, religion, theology, and spirituality. Moreover, grounded as it is in Maslow's "Hierarchy of Needs" (1943), this approach to nested ecology operates on the assumption that in order for an individual to address basic "social-ecology" needs, they must have satisfied their basic self-ecology needs. Likewise, realizing environmental ecology needs requires that basic social-ecology needs be fulfilled. A similar requirement holds for achieving a sustainable cosmic or spiritual ecology. However, it is of particular concern and interest that we consider more fully just what constitutes "basic" needs at each of these ecological levels. What are the basic needs that need to be satisfied for a person to feel as though they have successfully taken care of their basic housekeeping responsibilities at each ecological level?

SEVEN

The Fundamentals of Nested Ecological Householding

Basic Ecological Needs and Human Responsibilities

I have asserted throughout this text that a nested approach to understanding ecology is a practical and functional way of interacting within the world. I have also discussed these nested ecological domains in terms of health and sustainability, pragmatics, and hierarchy. Ultimately, however, a fully functional nested approach must fulfill a set of basic ecological needs and functions at each nested level that promotes the process of creating and maintaining a human household (i.e., householding) in a fashion that is responsive to the householding needs of other people, species, communities, ecosystems, and environs.

Obviously, creating a list of expected human values and behaviors that address an exhaustive set of needs particular to these nested ecologies would be tedious and time-consuming beyond the scope of this book. However, it is feasible to identify a set of *basic needs* indicative of responsive and responsible ecological householding at each nested level and to respond to these needs with values, expectations, and actions that promote homemaking and housekeeping (meant literally and figuratively) in a fashion that is responsive to the householding rights and needs of every other species, community, and ecosystem. This chapter will discuss basic needs specific to each nested domain and will then suggest a set of attitudes and behavioral functions that best serve to meet these needs. It is important to realize that since "nested ecology" is a new way to conceptualize ecology and human responsibility toward the environment, neither the listing of basic needs identified nor the suggested responses to these needs can be expected to be fully inclusive. Nevertheless they constitute a

146

starting point in the dialogue regarding what people need to minimally do to foster ecological sustainability at each of these levels, while insuring that basic human needs and desires are also met.

Personal Ecology

Basic Personal-Ecology Needs

Maslow's hierarchy provides a useful perspective on the basic ecological needs that people require if they are to achieve a sustainable personal ecology. These include meeting basic physiological needs such as a steady and satisfying source of food and nutrition, health, some degree of physical comfort, sufficient clothing and physical protection from the elements, and the like. Physiological needs represent the most basic of human necessities and occupy the base of Maslow's hierarchy-of-needs pyramid. Realizing human potential at any higher level is entirely dependent upon satiating these elementary needs. Alternatively, failing to meet some or all of these needs condemns people to ill health, unhappiness, personal and social disequilibrium and, in the most extreme cases, death.

Beyond these physiological needs lie what Maslow refers to as safety needs. They include not only being physically secure from human or environmental trauma and distress but additionally entail feeling psychologically and socially safe. The realization of safety needs includes all of those things that are necessary to vouchsafe the wholeness and integrity of a person's body and their personal sense of self. Assuming these basic safety needs are realized, then a person can pursue more complex interpersonal goals — chief among them being the pursuit of love, affection, and recognition. The realization of these goals, however, presupposes a social environment in which others can also readily realize their physiological and personal safety needs. If these basic needs cannot be met, then there is no chance that an individual will find another person who — having been freed from concerns at these most basic of human levels — will be available and capable of entering into intimate interpersonal relationships involving trust, caring, love, and loyalty.

Finding someone to engage in a successful loving relationship is one thing, but being able to integrate the affection and respect of others into one's own sense of self is a more involved and complex process. That is why Maslow identifies the realization of self-esteem as a higher order achievement. He does so because it entails psychologically internalizing the sense of self one derives from one's most loving and intimate relationships with others into an inwardly held sense of self that is generally consistent with how others regard them. This is what sociologists and social psychologists sometimes refer to as the social construction of self (Harter, 1999) since the internalized image of who one is and of one's value in the community is principally derived from interactions with others. For this interpersonal interaction to result in a sense of self-esteem, it is necessary for a person to have consistently entered into nurturing relationships with others and to have internalized the best perceptions of others regarding who they are into their personality.

Assuming all other lower-order needs are met, Maslow asserts that one may proceed to realizing "self-actualization" needs. These include freeing up time, energy, and resources to ask questions such as "Who am I?" "What is my purpose in living?" "What is my place in the world — where do I belong and who with?" "Am I realizing my full human potential?" "What will my life mean for others?" and "What is the meaning of life and death?" These questions and others are the perennial preoccupation of human beings and are the kinds of practical and philosophical questions people may ask and will inevitably pursue once more basic lower-order needs are satisfied.

Beyond "self-actualization" Maslow identifies the need for "self-transcendence." This embodies the need to connect to and feel a part of something larger, greater, or bigger than oneself (Maslow, 1943; Scheff, 2007). This need is the impulse that motivates people to engage in spiritual or religious activities (Maslow, 1943; Scheff, 2007). Moreover, it is likely the creative source for the arts, literature, and philosophy and is the principle factor behind the human need to grapple with the "ecology of the unknown."

Maslow's categorization of basic psychological needs remains germane and applicable to modern times and circumstances. However, they were formulated in 1943, and since then a prodigious volume of psychological research has been generated relative to understanding human needs and motivation. Recently this body of literature was reviewed and summarized by psychologists Thane Pittman and Kate Zeigler (2006), and the outcome of their work suggests some further refinement of Maslow's perspective on basic human needs at the psychological level.

Pittman and Zeigler reviewed the literature about human needs and categorized needs into three groups: (1) needs emanating from "basic biological-level processes," (2) needs emanating from "individual-level processes," and (3) needs reflecting processes at the "social process" level. It appears that at the level of "basic biological processes" the needs for food, water, clean air, and basic warmth are considered, as well as the ability to experience pleasure and pain. At the level of "individual processes" (the level of the individual person which assumes that an individual person can achieve goals independent of the community), can be found needs such as Maslow's self-esteem and self-actualization, the need to express innate curiosity, competence, and personal autonomy, as well as the realization of a sense of meaning and self enhancement. Finally, at the "social process level" (where human needs must be realized through interaction with others—though recognizing that many "individual process" needs can also be realized socially) research suggests that humans have needs for relatedness and belongingness, attachment, affiliation, caregiving (receiving and providing care), communal learning, and for shared culture (Pittman and Zeigler, 2006).

Essential Dimensions of Personal-Ecological Householding

While the inclusion of these need considerations does much to bolster Maslow's paradigm of human needs and motivations, I don't think the research quite captures the "householding" needs of individuals—which is to say that it doesn't contextualize basic needs against the essential task of making a home and caring for household needs at each of the nested ecological levels. More specifically, basic needs have been identified and articulated from a perspective other than the unique one involved in making and maintaining a household. Such a personal perspective, to be successful, must maximize the health of the individual and interpersonal relationships and should be sustainable over time. If one approaches identifying basic ecological needs from a personal householding perspective, then the categorization of such needs must functionally address issues associated with *making a home* (securing, provisioning, and maintaining a safe and healthful living space) and *being at home* (experiencing a sense of belonging in a particular place, locale and time, feeling at home with one's family, friends, and neighbors, and feeling at home with oneself).

Making and Maintaining a Home

Acquiring a healthy and sustainable ecology of self entails locating oneself in a safe, sanitary place where one can reliably experience comfort and protection from the elements while situating oneself within reasonable proximity to a set of basic resources (food, water, fuel, work, community). Ultimately it needs to be a site that the dweller comes to regard as his or her "place," and one that leaves the person with a sense of belonging and permanence — even if it turns out to be a comparatively temporary home.

Personal ecologies are also dependent upon developing the ongoing capacity — physically, economically, and cognitively — to continually provision one's home with basic resources as well as to acquire those additional provisions that allow one to comfortably and safely reside within the home. This entails developing the capacity to physically maintain a clean and sanitary home. Failure to adequately provision and maintain such a domicile makes it impossible for people to sustain themselves as individual persons (selves) within the larger world.

Finally, personal ecology also demands that one treat one's body as a biological home — caring for and nurturing it so that it can continue to serve as the physical embodiment of one's personality and self. This requires attending to dietary, hygiene, exercise, and physical and mental health needs. It also entails refraining from abusing one's body with foreign or harmful substances or agents as well as safeguarding one's body from physical or emotional abuse from others. Essentially, maintaining personal health is the biological equivalent of provisioning and maintaining one's physical home and constitutes "householding" at the physiological level.

Being at Home

Being at home implies that where one resides in place and time comes to be experienced as the place where one belongs and where one derives a sense of contentment, purpose, and meaning beyond the basic needs for safety and sustenance. Being at home applies not simply to the particular abode one lives in but also involves one's sense of attachment and belonging to the surrounding community and environment. It implies coming to personally identify with a particular place or locale, as well as suggesting that ultimately one's home will reflect—in terms of one's belongings and the décor of the surroundings—something about one's unique personality. While the home as a physical site or locale is not identical to its inhabitant or inhabitants, it is descriptive of the resident's personality, history, interests, and proclivities. Truly "being at home" is to experience a deep sense of place and belonging and is one of the hallmarks of a healthy and sustainable personal ecology.

Being at home in one's surroundings further implies considering natural environments surrounding one's abode as neighboring "households" upon which one's own household is dependent. Relating to these surrounding ecosystems in this fashion does not imply that one necessarily owns them or takes proprietary license to treat them as real estate. Rather it implies relating to surrounding natural environments as if they had an autonomy and complementary purpose independent of one's own and to judiciously interact with them, only drawing upon their resources as need dictates, and even then doing so in such a way as to minimally violate their integrity. Assuming such a perspective implies that one will minimize harm to these surrounding ecological households—recognizing of course that some degree of pollution is unavoidable for humans as a biological species and culture. Likewise, it implies that one will prudently draw upon the resources of these environs—always seeking to do so in a fashion that insures the ongoing sustainability of as many natural resources and life forms as possible. Persons who choose to ignore their dependence upon surrounding natural environments and systems do so at the risk of failing to achieve and maintain a sustainable sense of self and self-identity.

By definition, personal ecology involves not only the projection of "self" into the world within the context of a particular locale, place, and time, it also implies asserting oneself within the context of one's family, friends, and other social relationships and groups. In reality, the very process of acquiring a sense of self is fully dependent upon the others within one's network of family and friends. Naturally the most important of these include one's parents, siblings, extended family members, and family friends. A person learns who they are by gazing into the mirror of others' faces, listening to their voices, ideas, and opinions, and experiencing their behavior and ideas. These are the sociobiological forces that shape human identity. A person possessing a healthy and sustainable ecology of self has learned from these interactions that they are loved, valued, worthy, competent, and included within the family system and other social networks. Personal ecology also suggests that such a person makes good choices in terms of creating relationships with friends and intimates and that these relationships support and nurture the person's needs and aspirations. Ultimately, these relationships contribute to experiencing a "sense of place" within one's familial and social relationships that is akin to experiencing a sense of place within a particular locale or environment. Individuals who fail to assert a clear sense of self within the context of their families and others in their social network find it difficult to sustain relationships and identities throughout their lifetimes.

Ultimately the psychological and biological corollary of being physically at home in one's abode, family, community, and surrounding environment is coming to feel at home inside one's skin — at home with oneself. As Maslow observes, achieving such a sense of comfort with oneself is a higher order task that presupposes that a whole array of more basic needs have been realized. However, at the personal-ecology level, being at peace and at home with oneself is one of the most important of all needs since — in its absence — the realization of all other basic needs leaves one feeling ultimately unfulfilled.

Social Ecology

Basic Social-Ecology Needs: The Ascendency of Home Economics

Upon reviewing these basic personal-ecology needs, it is obvious that achieving virtually any of them presupposes a social context and requires a supportive social ecology. The two realms are intertwined (as is the case with every other nested ecological level). Since personal ecology emerged as a psychological phenomenon from social and developmental psychology, it tends to be construed as involving a set of psychobiological and psychosocial needs, and the paradigm that has been suggested for both conceptualizing and assessing these needs emanates from developmental and social psychology (Maslow, 1943; Pittman and Zeigler, 2006). Social ecology, by comparison, is best conceived of as a sociological, anthropological, economic, and political phenomenon. Given this orientation, basic social-ecology needs are best understood by using concepts from the social sciences.

Unfortunately, social ecology does not have an intellectual icon like Maslow to readily suggest what is entailed in meeting basic social needs. However, unlike personal or self-ecology, social ecology demands a virtually literal interpretation of the term "ecology" from the original Greek along the lines suggested by Ernst Haeckel. Haeckel's term in the original Greek (*eco-logos*) literally translates into "home knowledge," "knowledge of home," or "home economics." I personally like the last definition, despite its traditionally pejorative academic reputation of being a "soft" discipline in the academy, reserved for women attending college to gain an education and find a husband (Lang, 2000). Historically, home economics was a major reserved for those who were perceived to be less intellectually motivated (regardless of whether they were or not) and was an academic discipline widely and chauvinistically considered to be a "woman's" major (Yaffe, 2005; Laughlin, 2000; Stage and Vincenti, 1997).

I find it truly ironic that this underappreciated and undervalued discipline — home economics — has become so central to the modern quest for ecological survival. In fact, I assert that home economics has become *the very discipline* around which future communities, culture, government, and values must be shaped if we are to survive as a species and transfer natural and productive environments to future generations. Home economics (i.e., managing communities as households — personal, social, and environmental) has become the central task of humankind in an era of ecological crisis. No longer is it a devalued discipline relegated to those lacking ambition or intellect. It is now the principal task of all people, regardless of their economic or personal characteristics, and is an endeavor upon which the fate of humanity and the natural world as we know it depends.

Home economics provides a unique perspective on social ecology since it prioritizes the basic human need to belong and form meaningful social attachments and contextualizes this primary need at the level of the family and the family's relationship to the individual and the larger community. Available research suggests that the need to belong is a fundamental and pervasive human motivator that may serve as a primary impulse through which all other needs are contextualized, considered, and pursued (Baumeister and Leary, 1995). Accordingly, research findings suggest that people readily form social attachments even under the most stressful of situations and seek to maintain existing attachments. These social attachments exert a variety of strong effects on human emotions and cognition, and when they are fractured or missing, deleterious physical and mental health effects can ensue — effects that can be reversed when attachments are reinstated.

I likewise appreciate home economics' emphasis upon the family as the unit of analysis, since I consider it to be the social corollary of looking at personal ecology from the perspective of individuals and their biological and psychological needs. I favor approaching social ecology from the perspective of home economics because it is principally concerned with managing family households (the "householding" of this book). Home economics becomes the logical paradigm of choice for considering social ecology, and the family becomes the logical unit of social ecological reference and analysis.

Despite my predilection for home economics as an orientation to social ecology, I recognize contributions that the rest of the social sciences make toward realizing a sustainable social ecology. For instance, from a political-economic perspective, the United Nations has proffered a list of what it calls "basic human needs" based upon a set of international surveys and studies. These needs include the right to work; the right to fair wages; social security, including social insurance; an adequate standard of living and the continuous improvement of living standards; the highest attainable standard of physical and mental health; access to education, including free and compulsory primary education; the benefits of scientific progress and its applications; and an opportunity to participate in cultural activities (UN, 1998). For each of these "basic needs" the United Nations asserts a set of "basic rights" — thereby advocating for a set of basic living conditions and social expectations that all nations and societies are expected to achieve.

The basic rights proposed by the United Nations are similar to another set of basic requirements proposed by the Canada International Development Agency (CIDA, 1997) in their "Policy on Meeting Basic Human Needs." This policy, which is based upon the 1995 report of the World Summit for Social Development, calls for access to preventive and tertiary health services, universal literacy, and a minimum of a secondary education, full access to family planning and reproductive health services, basic nutritional needs, safe drinking water, sanitary living and work conditions, and access to safe and secure shelter for all people. Likewise,

CIDA chooses to reformulate basic needs into an array of basic services to which a person may or may not be entitled.

Basic human needs are articulated somewhat differently by sociologists. For example, Sandra Marker (2003) construes basic needs as including safety/security, belongingness/love, self-esteem, personal fulfillment, self-identity, cultural security, personal freedom, and social participation. Sociologists have also considered basic human needs from the perspective of what some refer to as "human flourishing"—the pursuit of human happiness (Paul, 1999). In this regard, basic needs have included insuring that the material necessities of life are securely met; decreasing the time people spend in toilsome labor, increasing people's control over their own lives; guaranteeing constitutional rights; bridging social inequities of race, gender, and class; and insuring fair compensation for loss of income from natural disabilities, or acts of nature, to identify but a few. Meanwhile, social workers such as Dean Hepworth and Jo Ann Larsen argue for a set of basic needs comparable to those of the UN and CIDA, including an expansive set of services ranging from social support to health, recreation and income support services (Hepworth and Larsen, 1993).

While many of these basic needs are unquestionably applicable to social ecology and may ultimately be included in the home-economics paradigm I propose, most fail to approach social ecology from the perspective of the family and the basic household needs families require to minimally function in a human community. In truth, I wish that I had been the first to associate the virtues of home economics with ecology, but I was not. That distinction resides with eco-philosopher Wendell Berry and his 1987 book *Home Economics*—considered by many to be his most important work. However, Berry's pragmatic and practical outlook on ecology has not been embraced to the extent that it might have, leaving writers such as myself with the task of articulating its promise anew—this time regarding its applicability to social ecology.

From Home Economics to Family Ecology

In recent years, the field of home economics has gone by any number of names including consumer sciences, human ecology, and family ecology. Historically, home-economics programs were often involved in curriculum dealing with early childhood education and child development. These courses focused upon family relations and child rearing, as compared to more traditional home-economics emphases such as home maintenance and provisioning, garment making, gardening, and the like. Family ecology, as a discipline, appears to have integrated home economics' traditional emphasis upon parenting and child rearing.

Home economics in the form of family ecology is actually an ideal starting point for thinking about social ecology, since it principally focuses upon the family (as opposed to the self) as the basic unit of social analysis and seeks to promote family health and well-being in a sustainable fashion — i.e., over the long run and across family generations. As such, families are studied and approached as a set of nested social systems — identical to how ecologists approach natural systems. To that end, family ecology affirms an array of social-system characteristics that include interdependence, role/system differentiation and integration, circularity, adaptive childrearing, open communication, rulemaking, homeostasis, and morphogenesis (Minuchin, 1998; Connard and Novick, 1996; Stafford and Bayer, 1993; Fine, 1992; Krauss and Jacobs, 1990; Walsh, 1982).

While families serve as the basic unit of social ecology, they are also the building blocks upon which more complex social-ecology systems are created and bridged. Successful families must effectively facilitate interactions along a progressive network of nested interrelationships, including extra-family relationships with individuals and groups, informal social networks, communities and community systems, and the broader society and culture.

While family relationships begin with interactions among family members, they quickly extend to include interactions with members of other families and groups. Thereafter, they expand to include informal social networks of individuals organized beyond the family level. These social networks integrate the family into the larger community and serve as the principal vehicle through which family needs are met and social communication achieved. Furthermore, they serve to define the validity, worth, and identity of the family as a unit as well as of its individual members and aid the family in adjusting to changes and demands from the larger, external environment.

Families organized according to a complex set of family and formal and informal social systems comprise a community that in turn serves as the context for all familial and social discourse, commerce, sustenance, and identity. Communities and systems of communities nurture and reinforce family values within a given local geographic area. Communities also foster the development of an array of economic and community resources upon which families rely. Ultimately, communities become integrated together to create a rich network of "communities of communities," or "society" (Etzioni, 1998). These community networks serve to define and organize a set of smaller nested subsystems and provide the overall social, economic, legal, and political structure within which families and their constituent members reside, work, recreate, socialize, and serve. Ultimately society is a composite of all the smaller social systems and subsystems it incorporates and subsumes.

Within the bounds of this social-system paradigm, hierarchy is nested, beginning within the family and extending outward to create societies, with their formal and informal structures of authority. This insight is important for understanding social ecology since it implies that while the basic building block of social ecology is the family, social ecology also requires a broader understanding of systemic relationships at the level of informal family networks, the community, and the larger society.

Essential Familial Dimensions of Social Ecology: Needs and Functions

In fact, social ecology is a threefold construct. At the core is the family unit. In the absence of the family, communities would cease to exist. Communities, in turn, serve as the building blocks of society since, as Etzioni (1998) observes, societies are no more than "communities of communities." Ultimately society serves as the aggregate reflection of human existence and culture on the planet. It is consistent with the nature of human beings as biological creatures that they must necessarily project their existence in this threefold fashion as the social expression of their biological identity as a species. Such is the nature of their "necessary anthropocentrism."

The basic foundation upon which social ecology rests is the family. Most experts on families and family health assert that the primary purpose of a family is to support the growth, health, well-being, and personal enhancement of its members (McCubbin and Patterson, 1983). However, the process of supporting such growth and development entails creating and maintaining a household and home—not just in the concrete and physical sense of the term, but also in the symbolic, psychological, cultural, economic, and sociological sense. From the perspective of social ecology, creating a family household minimally entails insuring that basic family needs and functions are sustainably realized.

Social ecology is a composite process that begins with satisfying basic needs at the family household level and thereafter at the level of the larger community or communities within which the family unit resides. At minimum, it involves creating a healthy and sustainable family household capable of addressing basic family and community needs and requirements. Given the current state of knowledge and expertise regarding families emerging from psychology and the social and behavioral sciences (Epstein et al., 1993; Schweitzer, 1993; Curran, 1983; Beavers, 1982; Walsh, 1982; Lewis, 1979; Stinnett, 1979; Otto, 1962), a sustainable and healthy family system is expected to functionally provide for a variety of basic needs involving leadership and support (household leadership, family nurture, and member support), procreation and parenting (conception, childbirth, and child rearing), identity and interpersonal interaction (as reflected in member individuation, autonomy, growth and independence, interpersonal intimacy, family belongingness, and acceptance), and spiritual nurture.

Leadership and Support / Procreation and Parenting

All families require leadership, either in the form of a married heterosexual couple (husband and wife; parents), a single parent or person, a same-sex couple (with or without children), or extended family members, including grandparents, aunts, uncles, etc. The function of family leaders is to provide organization, rules, guidance, discipline, and order among family members and to provide the requisite economic, social, and psychosocial support that family members require to be sustained and grow within a family. Ultimately family leadership should enable and facilitate the process of members achieving individuation and emancipation from their families of origin so that they might form new family systems in which they assume leadership roles.

Families also serve an integral procreative and child rearing function and are the foundational social systems for human physical and psychosocial development. Healthy and sustainable families are those that engender and facilitate successful procreation and child rearing in nurturing environments where all of the child's needs for physical, social, educational, moral, and emotional maturation are met. If these basic needs are not facilitated and realized, then families and, ultimately, communities cease to exist and function.

Identity and Interpersonal Interaction

Sustainable and healthy families serve as the most essential and foundational reference point for family nurture, love, identity, and support. Individuals acquire their most basic personal identities from within families and are ill-equipped to function either within or beyond the bounds of the family unless they have acquired a positive sense of themselves and others as reflected in the nurture and love they have experienced within their families. Families that fail to provide a basic level of nurture and social support, and that fail to teach their members to extend nurture and support to others, produce children incapable of effectively and satisfactorily functioning at any other higher social level.

Family members, while sharing a common kinship, are nevertheless unique individuals and require family systems that allow and encourage them to individuate. Realizing such individuation requires a family environment that is sensitive and responsive to the divergent needs of family members. Healthy and sustainable families are those which discriminate between the variety of needs of family members and seek to insure a degree of balance among members in the interest of insuring that, within the bounds of family resources, as many of each individual's needs as possible are met.

Such families engender a sense of intimacy between members that serves as the context for the giving and receiving of love, affection, and nurture. These qualities, in turn, become internalized into the personalities of each family member, providing the sense of self worth and meaning that is necessary for individual members to ultimately individuate, become autonomous and independent of their families of origin, and form new family units. Families must likewise be able to effectively engender a sense of acceptance and belongingness within the bounds of the family system. This basic sense of acceptance and identity is essential for the psychosocial development and individuation of self-assured, confident, competent, empathetic, loving, and secure people.

Spiritual Nurture

Healthy and sustainable families define themselves across generations by developing rituals and practices dealing with birth and death in their households. Central to this effort, and in keeping with the need for families to develop a sense of purpose, identity, and mission, is the need for families to articulate a spiritual vision of the family and the family's place in the world that transcends existence. Spiritual nurture enables families to cope with the vicissitudes of life and the inevitability of death. Such spiritual nurture may occur entirely and even informally within the bounds of the family or may extend beyond the family — integrating family members into a larger religious community or belief system. Regardless of how this occurs, successful families prepare their members for coping with the hardships and finitude of life by providing them with a spiritual perspective that places each member and their family within the context of a much larger and longer continuum of life and death.

Basic Family Householding Functions

Beyond meeting basic family household needs, sustainable and healthy family systems also promote a set of basic household functions and activities, including maintenance and integration of system boundaries (defining and maintaining the family's unique identity, purpose and mission; integrating the family unit in a meaningful way into the larger community; and integrating the household into natural environments), system communication (promoting clear and honest communication patterns and skills; conveying family values, identity and mission) and procurement and management of system resources.

Maintenance and Integration of System Boundaries

Intact and fully functional families define and enforce a variety of boundaries requisite to maintaining family roles and relationships as well as defining the relationships between families and other extra-family systems. These boundaries include the social boundaries that delineate the roles and expectations of parents and children, spousal roles, and the complementary roles of extended family members. External boundary creation and maintenance defines and protects family members from unwarranted intrusion from extra-familial sources as well as aiding members in understanding what distinguishes the unique identity and values of the family as a social unit from those of other families, groups, and social systems beyond the boundaries of the family.

Families do not exist in isolation from the larger community and, to be healthy and sustainable, must involve themselves in extra-familial activities and relationships that serve to define the family and its members as members of a larger social community. No family can fruitfully function in isolation from other families and communities. Family and individual meaning and purpose are necessarily derived from relationships between the family and the larger community. Fully functional families integrate themselves into the larger community and systematically organize themselves to insure that all resources — physical, economic, social, and spiritual — necessary to the function of the family as a unit are readily accessible and appropriate to family needs. Families who experience difficulties in achieving these basic needs find it very difficult to achieve any of the other basic needs and functions of the family. Additionally, families must integrate themselves into the natural environments that sustain and house them. The lifestyle, livelihoods, and identities of families and their members are as dependent upon how successfully and sustainably they integrate themselves into these natural systems as they are upon social-system integration.

System Communication

Families must engage in clear and unambiguous communication with their constituent members and engender the capacities and skills within family members that allow them to relay their own personal needs to others and to realize the satisfaction of these needs. The development of basic and effective communication skills is also necessary for effectively communicating and negotiating with extra-familial individuals, groups, and systems on behalf of the needs of the family and self.

Healthy and sustainable families utilize their system boundaries to establish a clear sense of identity and purpose, thereby allowing them to communicate what the family believes about itself and for what purpose it exists. Just as psychologically healthy individuals require a sense of purpose and meaning in their lives (Maslow, 1943; Cooley, 2004), so do families require a sense of purpose and mission to direct them as a unit and to integrate their identity into that of the larger community (Becker, 2005; Leider, 2005; Covey and Covey, 1998; Schorr, 1998).The development of family purpose and mission is essential to the development of family values, which are later conveyed beyond the boundaries of the family to become social values, mores, ethical codes, and laws.

Procurement and Management of System Resources

Viable families must establish a sustainable relationship with the natural environments in which they are situated. Families whose demands and desires outstrip the resources of the natural environments that support and sustain them not only run the risk of permanently destroying these environments, they run the risk of destroying themselves in the process. Successful families establish a mutual and complementary relationship with the natural environs they rely upon and ideally do so not simply on the basis of utilitarian, economic, or even aesthetic grounds, but out of a genuine regard for the independence and autonomy of nature as a "household" comparable to the deference families would want to see demonstrated toward their own family household.

Essential Community Dimensions of Social Ecology

Communities are of necessity reflections of the families that create, inform, and sustain them. Sustainable and functional communities presuppose sustainable family constituents. Consequently, the basic needs and functions necessary for family sustainability described above serve as prerequisites for the realization of sustainable communities. However, at the community level, extensive research suggests that there are a number of attributes and functions that must additionally be realized to achieve sustainability (Kessler, 2003; Wolff, 2003; Anderson, Shinn, and St. Charles, 2000; Berkowitz and Cashman, 2000; Norris and Howell, 1999; Duhl and Hancock, 1997; Adams, 1995; Flower, 1993; Schweitzer, 1993), including those involving community perspective and involvement (a shared and systemic perspective regarding community identity, vision, and purpose as well as a commitment to empowered citizen involvement, cooperation, and collaboration), community communication and conflict resolution (entailing effective interpersonal and intergroup networking, effective institutions and processes for conflict mediation, community continuity, and a complementary hierarchal system), and community and environmental sustainability (a commitment to develop the capacity for achieving sustainable human communities that also sustain natural ecosystems and environments).

Community Perspective and Involvement

Fully functional communities share a common historical perspective as well as a unified vision and sense of community purpose. The community's historical and current vision and purpose are in turn consistent with the divergent makeup and perspectives of the families that reside within the community and that serve as its primary constituents and leaders. The community's sense of vision and purpose influences family identity and purpose and is also a reflection of underlying family values.

Additionally, responsive communities affirmatively act to empower citizens, as members of family and community systems, to assume civic leadership and housekeeping responsibilities, thereby fostering a positive civic culture that promotes participation and responsibility on the basis of a practical concern for the greater needs of the community and its constituent families and individuals. Such communities engender citizen cooperation and collaboration while minimizing special interest entrenchment, polarization, and discord. To that end, groups and individuals seek compromises that promote community consensus and teamwork.

Community Communication and Conflict Resolution

Intact and lasting communities consist of effective networks of communication and cooperation from both within and across the community itself, as well as across and among other neighboring communities. Since communities are hierarchically organized into larger "communities of communities," effective networks of communication and cooperation must be maintained vertically as well as horizontally across communities. Effective community communication networks must be accessible to the family systems that are the foundations of the community to insure that their needs and values are incorporated into community values, activities, and functions.

In a similar way, sustainable and fully functional communities foster those basic foundational organizations around which community needs, governance, and services are developed and delivered. These institutions seek to develop effective mediating organizations, structures, and systems that ameliorate and avoid unnecessary community conflict. These are the foundational structures and organizations in which citizens from diverse community groups and backgrounds can collaborate to arrive at mutually acceptable decisions about the community and its future and avoid unnecessary and fractionalizing conflict.

Communities that promote such institutions eschew authoritarian social structures where power emanates at the top and is directed downward into the community. This is the authority structure that social ecologists like Murray Bookchin (2003, 1986) particularly despise. Comparatively, within a complementary hierarchy, authority patterns are reciprocal, emanating from below (among consensus makers in the community), from within the midlevel regions of professionals and institutional bureaucratic officials and advisors, as well as from above — the highest levels of formal and informal leadership and influence. As a result, authority legitimizing action and policymaking is a product of the interaction of these various levels and sources of authority (citizen consensus, professional, and upper leadership), insuring the widest possible level of participation in community decisions.

Community and Environmental Sustainability

Successful communities also develop the capacity to anticipate, mediate, and react to problems and issues as they arise — including anticipating issues associated with the flow of basic resources into the community — resources necessary to sustain community functions. Ideally, healthy and sustainable communities plan for and solve problems and issues on the basis of the assets the community possesses or commands. This is particularly the case when community problem-solving strategies tap into capital or fiscal resources beyond the community's ability to repay or replenish, or overutilize natural resources that are nonrenewable or cannot readily be augmented, replaced, or substituted with renewable resources.

Capacity building is absolutely essential to the health and viability of communities. Capacity building implies assessing existing community strengths and weaknesses and on the basis of this inventory investing in and enhancing the community's "civic infrastructure." As was the case for problem solving, capacity building includes developing the resources necessary to build and maintain community capacity and provide for community needs. These activities must occur within the available economic, ecological, and human resources of the community. Community resource demands that exceed resource capacity can not only financially bankrupt a community and/or overextend its human resources, they may additionally result in ecological overextension and invite environmental disaster. Without careful planning, capacity building can generate a level of community affluence and comfort that may prove to be short-lived. Once resources have been exhausted, the community may become ecologically and fiscally depressed, followed by social disintegration and/or the complete or partial dissolution of the community itself.

Functional communities must also insure an ongoing sense of community continuity at many levels, including social and political, resource, fiscal and economic, and cultural stability and continuity. Stable and predictable communities promote the interests of families, allowing them to function comfortably in the present and effectively plan for the future. Moreover, they provide the social stability required to live life fully, without undue worry or concern. Finally, community continuity insures that community values, ethics, policies, and planning extends into the future so that the community can assert its values from the past into the present and forward into the future.

Finally, healthy and sustainable communities only exist if they are supported by vital and fully functioning environmental ecosystems. These natural environments provide all of the wherewithal — water, food, energy, and the like — upon which human communities are dependent. They also provide an aesthetic value and context for communities, rendering them unique and pleasing to the senses while engendering a special quality that contributes to a community's environmental identity and sense of place. Beyond these utilitarian considerations, healthy and sustainable communities will recognize environmental ecologies as consisting of autonomous "households" that possess an innate right to exist and function independent of human need or considerations. Human communities will seek to sustain and protect these natural environments out of deference to their deeply held civic, ethical, spiritual, and religious values.

Environmental Ecology

Basic Environmental-Ecology Needs

As the self serves as the basic unit of analysis for personal ecology, and the family serves as the basic unit for social ecology, so the ecosystem serves as the basic unit for environmental ecology. In the previous chapter we discussed what is required for an ecosystem to be considered healthy and sustainable (Muñoz-Erickson, Aguilar-González, and Sisk, 2007; Rapport, 1998a, 1998b; Arrow et al., 1995; Costanza 1992). Based upon this characterization, basic ecosystem needs could be arguably construed as minimally including vigor, organization, resilience, stability, constancy, diversity, and sustainability. Ecosystems failing to meet these minimum needs are simply unviable.

Ecosystem Vigor

Vigor is a measure of how an ecosystem functions over time (Mageau, Costanza, and Ulanowicz, 1995). Moreover, it represents the sum of all individual exchanges occurring within an ecosystem (Pimentel, Westra, and Noss, 2000). So conceived, vigor is an important component of ecosystem health as represented by the formula $HI = EV \times O \times R$ (Costanza, 1992) in which the health index (HI) of the ecosystem is represented as the product of ecosystem vigor (EV, denoting system activity, metabolism, and productivity), organization (O, or degree of system incorporation, including diversity and connectivity), and resilience (R, or the degree of rebound a system has following disturbance or insult).

This formula reflects another principal characteristic of ecosystems, namely, their reliance upon natural disturbances such as fire, wind, or flooding, for system invigoration. Consequently, not all ecosystem disturbances can be considered harmful (Johnson, 1994). This includes human disturbances as in those applied in agriculture and forestry. Vigor is also a function of the diversity of available natural resources, biota, and soil, as well as nutrients, water, and climate resources requisite for growth and change within the ecosystem. The more invigorated the ecosystem is, the greater the degree of interaction among life forms and the greater the degree of energy transformation and exchange.

The implications for maintaining the vigor of an ecosystem are clear. Maintaining the integrity and sustainability of an ecosystem *does not* necessitate protecting it from any or all disturbances. Rather it entails protecting it from those disturbances that do not substantially invigorate ecosystems, that deplete the diversity of natural resources and biota characterizing ecosystems in their "typical" state, and that dramatically transform ecosystem organization into an unnaturally altered state. To that end, promoting ecosystem vigor entails judiciously interacting with and using its resources within the bounds of its resilience and organizational capacities.

Ecosystem Organization

Ecosystems consist of an array of communities as well as the surrounding abiotic environment (the soil, water, elevation, climate, etc.). Communities are progressively organized from species to populations of species and may be either terrestrial or aquatic in form. They may exhibit characteristics of closed systems, in which species within the community share a comparable geographic range and density, or open systems, with species within the community distributed in a random fashion (Begon, Harper, and Townsend, 2000).

Ecosystem composition and distribution patterns are unavoidably linked to the diversity of biotic communities and their constituent species populations. Generally speaking, the greater the organizational complexity of the ecosystem, the more stable it becomes. Lower levels of ecosystem organization complexity may portend less overall stability by lessening aggregate species diversity (Nunes et al., 2003; Naeem, 1994). Diversity loss in an ecosystem also serves to weaken the resilience of the system, increase the variability and decrease the effectiveness of environmental processes such as those involved in soil nitrogen levels, plant yield, and water use. Decreased biodiversity can also be reflected in the growth of pest populations and increases in the cycle of diseases (Naeem et al., 1999; Perrings, 1996; Bakkes, 1994).

What this means in terms of human interaction with ecosystems is that environmental disturbances caused by the human species and their communities should not be so extensive as to significantly compromise the structural integrity and complexity of ecosystems. Such extensive disturbances render systems less complex and, by extension, less stable. They impair the sustainability and health of the environment for human communities (a narrowly utilitarian consideration) and for other nonhuman species and communities. Once again, while it is unreasonable to expect human beings to not disturb the ecosystems they share with other species — and in fact ecosystem invigoration is directly related to being disturbed (even by humans) — the level of disturbance should not be so extensive as to significantly reduce ecosystem diversity and stability.

Similarly, since ecosystems also consist of the abiotic environment, human intrusion and disturbance should not result in radical changes to the elemental composition of the environment (Morin, 1999). For instance, if wetlands are to remain productive for their constituent communities, adequate water resources must be maintained and few human-induced drainage or elevation activities introduced. If the ecosystem in question is in an arid region, then care should be taken to not introduce too much moisture. Admittedly, there will be instances when it will be necessary to drain and elevate a wetland ecosystem or to hydrate an arid one to meet human needs, but ethically doing so within the parameters of this nested-ecology approach would require those initiating such change to mitigate their impacts on any particular ecosystem by restoring a despoiled ecosystem elsewhere (Salvesen, Marsh, and Porter, 1996).

Ecosystem Resilience

As previously noted, resilience involves the degree and types of perturbations an ecosystem can absorb before the system is destabilized and transformed from one state into another (van Nes et al., 2007; Holling, 1996; Holling, 1995). Resilience can also be construed in terms of temporal elasticity — the time it takes to rebound — following a disturbance (Pimm, 1991). Resilience primarily pertains to overall ecosystem functioning and, to a lesser degree, reflects subpopulation stability or homeostasis (Gunderson and Holling, 2002; Gunderson et al., 1997; Gunderson, Holling, and Light, 1995; Holling et al., 1995; Naemm et al., 1994; Pimm, 1984). Resilience may also serve to stabilize ecosystem function and promote sustainability (Mooney and Ehrlich, 1997; Tilman, 1999, 1997; Common, 1995; Schulze and Mooney, 1993).

Not surprisingly, given the variable's complexity and the difficulties associated with directly observing it, measuring resilience has proved to be difficult, resulting in a wide variety of indicators being utilized as proxies for changes in ecosystem resilience (Tilman, 1999, 1997; Pimm, 1984), including changes in vegetation, invasive plant infestation rates, post-perturbation regeneration, species abundance and diversity, air and water quality, and ecological "rebound" following perturbation. Some of these indicators involve species and communities while others involve key ecosystem processes and functions. Generally speaking, individual species and their communities appear to be more sensitive to perturbations within ecosystems than are ecosystem processes themselves (Schindler, 1990). Changes in system resilience following disturbances and stress may be more obvious among indicator species and communities and less discernable within ecosystem processes and energy flows.

Further complicating the matter is the relationship between ecosystem redundancy (the assumption that various roles within an ecosystem are performed by a variety of species, such as in the case of butterflies and bees as pollinators) and resilience. Depending upon which study one relies upon, redundancy may or may not contribute to ecosystem resilience (Naeem, 1998; Walker, 1992). Resilience may also be influenced by the adaptability of the ecosystem, or it can be defined as "the capacity of the actors in a system to manage resilience" (Walker et al., 2006). For instance, increases in the adaptability of communities within one part of an ecosystem might promote resilience in this sector while actually diminishing system adaptability and resilience in another. Similarly, ecosystems may become sensitized to periodic perturbations and show resilience to these insults only to demonstrate a comparative inability to rebound from unusual shocks. Finally, system adaptability might also negatively influence resilience in instances where human intervention to increase the efficiency of an ecosystem or to increase resource output occurs — as in the use of monoculture forest plantations (Walker et al., 2006).

Resilience becomes a very important issue for humans, who will unavoidably "perturb" and marginally destabilize all natural environments. As we are a species among other species, such perturbation is natural and to be expected. However, if humans profoundly disturb such environments beyond their capacities to rebound, or if their efforts, such as those involved in agriculture and forestry, fail to actually invigorate these ecosystems and instead impoverish them, then it is not only ecosystem resilience that is diminished, it is also the social resilience of human communities dependent upon these ecosystems (Adger, 2000). Once these environments have been destabilized, the resultant ecosystem and social changes may or may not be reversible, or may be reversible only over extremely long human time scales (Carpenter, 2003, 2001; Scheffer et al., 2001). Admittedly this is a particularly utilitarian perspective on resilience. However, even though it does not narrowly address the worth of ecosystems independent of human valuing, it nevertheless presents a pragmatic concern for human beings as a species and community.

Ecosystem Diversity

Diversity has been associated with ecosystem vigor, resilience, and health (Tilman et al., 2001). Truly recognizing diversity's value entails appreciating its nuanced expression within ecosystems. At least three constituent forms of ecosystem diversity can be identified: functional diversity, influencing overall system performance; responsive diversity, which promotes system resilience; and genetic diversity, which allows for species adaptation to historical patterns of perturbation.

Functional diversity focuses upon the number of distinct functional groupings within an ecosystem and how these distinctive communities interact to influence the overall performance of the system (Walker et al., 2006). However, research suggests that it is not simply the quantity of species diversity that matters so much as it is the quality of the genetic diversity within an ecosystem (Madritch and Hunter, 2002). What this means is that it is important to maintain species communities distributed throughout the ecosystem that have genetically adapted to historical system disturbances. When this genetic diversity is reduced, then the integrity of the system may be permanently destabilized.

By comparison, responsive diversity pertains to the variety and modes that functional groups exhibit in response to perturbation, which is to say it studies those functional ecosystem patterns that ultimately produce resilience (Walker et al., 2006). This latter form of diversity is sometimes referred to as "functional redundancy" (Wohl, Arora, and Gladstone,

2004). Generally speaking, most ecosystems are assumed to exhibit a significant degree of functional redundancy—which is to say that they possess a variety of species and system functions that serve a common purpose. Consequently, vital ecosystem processes such as energy flow and nutrient infiltration and recycling are not believed to be seriously impaired by marginal species diminution, given the assumption that other species who function similarly to the displaced species will functionally replace them. Some ecologists prioritize the structural integrity of ecosystems over their functional properties—a proposition that remains controversial (Connell et al., 1999).

Human use of ecosystem resources principally operates upon the assumption of the system's inherent functional redundancy and diversity. While this may typically be a reasonable and sound supposition, given that human communities must perturb ecosystems and will unavoidably reduce species diversity to some degree, it should not be construed as a license to do so. Likewise, it would be unwise to automatically assume that every ecosystem possesses the same or even comparable capacities for functional and responsive diversity. Such is not the case, and proceeding upon this assumption can produce very serious deleterious effects upon ecosystems and the human communities relying upon them. A more reasoned and measured approach to human interaction with ecosystems is to seek to promote species diversity functionally, genetically, and responsively so as to vouchsafe the system's vigor and resilience over the long term, and with it the vigor and resilience of human communities.

Ecosystem Constancy and Stability

Ecosystem stability is conceptually similar to resilience and can be defined as the "continued existence of an ecological system or its capability to restore the original state after a change" (Gigon, 1983). In point of fact, some ecologists conceptualize stability in terms that are virtually identical with resilience. Certainly one of the most definitive comparisons of the two concepts comes from C. S. Holling, who observes: "Resilience determines the persistence of relationships within a system and is a measure of the ability of these systems to absorb changes of state variables, driving variables, and parameters, and still persist. In this definition resilience is the property of the system and persistence or probability of extinction the result. Stability, on the other hand, is the ability of a system to return to an equilibrium state after a temporary disturbance. The more rapidly it returns and with the least fluctuation, the more stable it is" (Holling, 1973, 17). Stability, as used by Holling, is virtually synonymous with elasticity (Light, 2001), and in that regard is very similar to the definition of resilience as temporal elasticity proposed by Pimm (1991).

Ecosystem stability can assume one of three forms: species stability ("the maintenance of viable populations or metapopulations of individual species"), structural stability ("the stability of various aspects of ecosystem structure, such as food-web organization or species numbers"), or process stability ("the stability of processes such as primary productivity and nutrient cycling") (Perry, 1995, 477). Given this typology, Perry asserts that

> the coin of stability has two sides and a middle: on one side are individual populations and metapopulations, which in the final analysis are the basic units of survival; on the other side are the higher-order aspects of system structure (e.g., guilds, communities, landscapes, regions, and ultimately the planet as a whole); the middle that ties these two together comprises . . . numerous relationships and processes . . . Successful conservation requires that we attend to the coin as a whole, and this in turn means protecting links that may be farflung in space and time. The multiplicity of both direct and indirect linkages that characterize nature raise the distinct possibility that to protect any one ecosystem, we must ultimately protect the integrity of the entire biosphere. (Perry, 1995, 477–78)

Perry's perspective is consistent with that of Vladimir Vernadsky (1926) as well as the nested-ecology thesis of this text. It presumes that the underlying ecosystem can be expected to generally remain stable and in this regard exhibit systemic resilience and resistance — a systemic ability "to absorb small perturbations and to prevent them from amplifying into large disturbances" (Perry, 1995, 478). However, it also assumes that the underlying state of the ecosystem (that second side to the coin that Perry describes) will remain constant.

Ecosystem constancy refers to a system that remains comparatively unchanged and permanent over time or one that "varies within defined bounds" (Belovsky, 2004, 99). Constancy, in turn, is vitally linked to stability such that if "*constancy* is a measure of an ecosystem's (in) variability through time and *stability* is a measure of the system's ability to damp and recover from environmental perturbations, then constancy depends not only on stability but also on the frequency and amplitude of perturbations — the *environmental 'noise level'* " (Crowley, 1977, 157). This is of course the critical factor to any ecosystem remaining stable and constant, since it is not a matter of whether ecological perturbations will occur but rather a function of their amplitude and frequency. Clearly ecosystems cannot control whether or not they are disturbed. However, they can be expected to have some capacity to constrain, absorb, or resist the degree of fluctuation they exhibit following perturbation. Again, as forest ecologist David Perry has observed, "a stable ecosystem is one that is capable of constraining its own fluctuations within certain bounds and, if not maintaining constancy in its species composition and productivity, at least maintaining constancy in certain potentials" (Perry, 1995, 476).

The ramifications for human communities regarding what is known about ecosystem stability and constancy is not that different than those we observed regarding ecosystem diversity and redundancy. As was the case for diversity and redundancy, human beings take for granted that ecosystems are stable and constant. Yet it is a given humans will legitimately threaten the stability and constancy of ecosystems via the process of making a home in these environs and utilizing their resources. The foreknowledge that most ecosystems are characteristically stable and constant should not lead us into assuming that they are endlessly so. Without question, a more reasonable option is to approach ecosystems with an eye toward how they respond to the cumulative effects of the perturbations of humans and other species as well as to external abiotic disturbances that similarly challenge the capacities of ecosystems. This means adjusting human-induced ecosystem disturbances to accommodate the combined volume of trauma from elsewhere and to constantly monitor the resilience and vigor of such systems to insure that human perturbances consistently occur at a "frequency and amplitude" (Crowley, 1977, 157) that the ecosystems can absorb and withstand.

Ecosystem Sustainability

In chapter 6 I discussed sustainability from an ecosystem perspective, including characterizing a sustainable ecosystem as one that consists of "ecosystem components and ecological processes that enable the land, water, and air to sustain life, be productive, and adapt to change" (Hammond et al., 1996, 2–18), or more simply in terms of systems demonstrating the capacity to remain healthy and thrive over the long run (Robbins, 1996). I summarized scientific literature suggesting that the factors contributing to ecosystem health likewise contribute to system sustainability. Consequently, ecosystem sustainability could further be construed as involving "ecological systems capable of maintaining organization, autonomy and resistance to stress" (Costanza et al., 1992) or systems that persist, maintain vigor, organization, and resilience (Costanza and Mageau, 2000).

However, succinctly defining what a "sustainable ecosystem" entails or, for that matter, suggesting what characteristics are indicative of such a system is extremely difficult precisely because sustainability is conceptually complex. Robert Costanza, however, asserts that an ecosystem should be deemed healthy "if it is stable and therefore sustainable" (Costanza, 1992, 7). In effect, Costanza suggests that a stable ecosystem is a sustainable one. While this may strike some as a gross oversimplification, his assertion is in fact more involved than one might think. Consider, if you will, Costanza's next words — those immediately following "if it is stable and therefore sustainable" — for they are illustrative of the complexity he perceives within ecosystem stability, namely, that such an ecosystem "is able to maintain its metabolic vigour, its internal organisation, structure and autonomy and is resilient to perturbations and stresses over a time and space frame relevant to the system" (Costanza, 1992, 7). These are the essential processes necessary to insure system stability and are likewise those basic processes needed to insure that the ecosystem persists over time in its essentially natural form. Costanza's sense of ecosystem stability's complex nature is also reflected in David Perry's (1995) taxonomy of stability, Holling's (1983) comparison of resilience and stability, and Naeem (1998) and Walker's (Walker et al., 2004) discussions of ecosystem redundancy, resilience, and stability.

Another perspective on ecosystem sustainability may prove to be most consistent with the nested-ecology approach of this work and may add a new dimension to understanding ecosystems beyond those previously discussed. This alternate perspective builds upon the fact that ecosystems are by definition open systems interacting with and exchanging energy with other open systems. Ecosystems are dynamic, nested, nonlinear systems that appear to self-organize — which is to say they evolve and emerge into higher-order network structures and systems (Schuster and Sigmund, 1980). Given these system characteristics and qualities, a sustainable ecosystem could be described as "a nested constellation of self-organizing dissipative processes and structures organized about a particular set of sources of high quality energy, materials and information, embedded in the environment" (Hearnshaw, Cullen, and Hughey, 2003, 15; Kay and Regier, 2000).

Such an involved description of a sustainable ecosystem is deserving of further examination. This energy-oriented perspective is based upon the "theory of dissipative structure" (Ziming, 1996) grounded in the Second Law of Thermodynamics (Fermi, 1956), which suggests that an open system, such as an ecosystem, exists in such a form that it can continuously import free energy from the immediate and surrounding environments while simultaneously exporting (dissipating) entropy — a measure of thermal energy unavailable for work, waste, or system disorder — across the system boundaries and into other systems. Accordingly, "order" in an open system can be maintained only in a state of disequilibrium, and to that end the system needs to continually rid itself of "disorder" or entropy. When such systems import high-energy resources from outside the system, they must simultaneously export low energy (entropy, waste, or disorder) from within the system. In so doing they maintain a state of flux or disequilibrium — a state essential to open systems, which require an ongoing exchange of energy and resources across the system boundary with the external environment (Anderson, 2005).

This ecosystem model recognizes that inevitably entropy (heat, disorder, and inefficiency) will accumulate in the open system and that order (or "negative entropy") is maintained by exporting this waste. Obviously one consequence of such an exchange is that the surrounding systems, having accumulated entropy, become more disorderly or chaotic — thereby contributing to what has come to be known as system chaos (Williams, 1997). If such open, dissipative systems are to maintain and sustain their growth and orderliness, they must regularly eliminate accumulations of system inefficiency, or, failing to do so, build up reserves of entropy sufficient to induce a state of thermodynamic equilibrium — a state in which energy transfer comes to a standstill (Harvey and Reed, 1996).

Based upon this scenario, sustainability must be accounted for in a new more informed fashion that captures the unique dynamics of ecosystems. Sustainable ecosystems are open systems consisting of nested constellations of self-organizing, dissipative processes and structures organized around high-quality energy resources and materials maintained in a functional state of system nonequilibrium — effectively exporting system inefficiency and disorder and importing energy. Given this definition of a sustainable ecosystem (one that exhibits vigor, resilience, stability, and constancy over time), several more specific considerations regarding sustainability come to mind. Sustainable ecosystems must first have access to energy resources sufficient to consistently meet their needs. If such resources are depleted, reduced, or eliminated, then the sustainability of the ecosystem is reduced or eliminated. Sustainable ecosystems must likewise maintain themselves as self-organizing systems — nonlinear, open systems in a state of disequilibrium, self-generating and renewing, steadily importing an abundant supply of energy and information from the surrounding environment. To remain sustainable, they must constantly move toward more orderly and predictable system functioning by exporting disorder beyond the system boundaries. Finally, having dispelled system entropy and disorder into surrounding environments and systems, ecosystems must maintain their capacities of resilience, constancy, and redundancy to withstand unanticipated disturbances from the chaos and disorder beyond the ecosystem — disorder and chaos that they have unavoidably contributed to by exporting entropy and that may ultimately threaten the integrity of the system.

Environmental Ecology's Co-constituted Householding Needs

These, then, are the seven basic needs of ecosystems — vigor, organization, resilience, diversity, constancy, stability, and sustainability. Ecosystems serve as the basic unit of analysis and function for environmental ecologies, and humans and their communities constitute but a small portion of these integral ecological units. Situated as we are within ecosystems and among a host of other plant and animal communities, we have the opportunity of regarding communities other than ours as neighboring households rather than ignoring their presence or narrowly relating to them as resources for our consumption and use. In every instance, nested ecologies — be they personal, social, environmental, or cosmic — are co-constituted among others, since there is simply no way that humans can fully shape or determine what "their" household will look like at any of these levels in isolation from the demands of neighboring households.

At the personal level the very identity of individuals is determined by their interactions with others and with the surrounding environment. This is even more so at the social level, as family identities are defined by their members as well as in interaction with the larger community. However, human beings have measurably more control in determining the shape and content of these first two nested households than they do in determining or even dictating what an environmental household will look like. At this level, more than at any other discussed so far, human beings share their personal and social households with — if you will — the personal and social households of countless numbers of other creatures whose right to existence in the larger environmental community is as valid as that of any person. Humans cannot as freely and readily assert a prerogative to determine what this ecology — this household — will look like, certainly not to the extent or degree that they do in shaping personal and social ecologies. At the level of environmental ecology, co-constituting the environment requires human beings to be cognizant of the vast array of ecosystems that the larger environmental household consists of and to assert the human influence in determining the shape and form of this larger environmental ecosystem in collaboration and harmony with the myriad other households and their inhabitants they share the world with.

Householding at the level of environmental ecology is discussed here from the human perspective in full recognition that countless other biotic forms also "household" — create and maintain a home or niche — without any assistance from humans whatsoever. However, since this treatise is written to assist humans in their interactions across a variety of nested ecological domains, it concerns how they co-constitute their households amidst those of other creatures and vouchsafe the integrity of the larger array of ecosystems upon which all creatures rely. Doing so requires us to formulate basic human householding needs within this ecological realm according to the basic ecosystem characteristics previously discussed.

Essential Dimensions of Environmental Ecology

Assuming that humans interact with their environments in a fashion that is sensitive to basic ecosystem needs, a functional, healthy and sustainable approach to environmental householding minimally requires that humans accommodate the householding needs of neighboring ecosystems and their inhabitants, particularly in regard to the functions of homemaking, community development, and provisioning and maintaining their needs with resources drawn from surrounding and distant natural environments. Ultimately, however, sustainable human householding requires that human beings come to "be at home," to realize a "sense of place" amidst nature, conceiving of their domiciles, families, and communities as residing in a more expansive set of natural environs. Rather than considering human communities and human beings themselves as somehow "unnatural" or artificial, a more eco-friendly stance would involve "householding" in which human households are viewed as akin to rather than distinct from the households of other species and the natural ecosystems that all creatures — including humans — inhabit.

Homemaking

Being at home in natural environments begins with the process of securing a home site. Doing so entails physically situating families and individuals in natural environments where they will utilize available resources to build homes and neighborhoods that minimally interfere with or transform local or distant ecosystems or stress them beyond their capacities for resilience, stability, productivity, or sustainability. It also involves minimizing pollution on home sites and committing to the mitigation of otherwise despoiled or impoverished ecosystems to as great a degree as feasible to compensate for those decrements in ecosystem function and capacity that have been occasioned by physically situating homes within natural environs.

In considering homemaking, it is important to recognize that most people — though certainly not all — have options regarding where they choose to physically locate their homes. When confronted with choices between ecologically sensitive and less-sensitive home sites, one would hope that they would choose those sites where building a home would place a smaller degree of stress upon surrounding ecosystems. However, the choice of where to make a home is typically defined by social and economic factors, such as where one grows up, derives a livelihood, or chooses to economically and comfortably reside. These considerations, along with those related to transportation, proximity to necessary community resources and services and access to cultural and recreational resources, often place people and families in environs where they would ideally prefer not to reside. In such instances, an appropriate ethical response at the environmental ecology level would be to seek to minimize human intrusiveness in surrounding environs, even if doing so requires families to live in less desirable areas, to expend more resources to make and maintain a home, or to constrain the scope of their family lifestyle in response to ecological considerations.

Community Development

As home after home is successfully situated, human communities emerge. Developing such communities entails geographically, socially, economically, and culturally locating human families and their households in natural environments that will provide them with the requisite food and resources necessary for establishing, sustaining, and growing communities. Ecologically establishing these households and their accompanying social, economic, and cultural supports, processes, and institutions additionally entails developing human communities within the bounds of what surrounding ecosystems can ecologically tolerate. This requires vigilance regarding the extent to which human communities can be introduced without critically compromising the vigor, integrity, and stability of nearby ecosystems beyond their capacity for resilience or to a degree that dramatically interferes with their long-term sustainability. It also entails minimizing pollution within and around community sites and committing to mitigate otherwise despoiled or impoverished ecosystems to as great a degree as feasible in order to compensate for those decrements in ecosystem function and capacity that have been occasioned by physically situating communities in particular environments.

Communities often have a great deal of discretion regarding where they situate themselves and the extent to which they will expand and develop their scope of influence and ecological involvement. Responsible communities, like responsible family households, will seek to minimize their ecological influence and imprint upon the surrounding ecosystems by efficiently and effectively organizing themselves to meet community needs without unnecessarily impinging upon the rights of surrounding natural ecological households. They will particularly focus upon delimiting population parameters and scope within the reasonable capacity of surrounding and even distant environs to sustain and support them.

Provisioning and Maintaining

Provisioning and maintaining homes, neighborhoods, and communities demands constant energy and resource input to produce needed goods and services — a process that also produces a steady stream of waste (entropy) that must be dispelled from human communities. Sustainably provisioning and maintaining human communities entails committing to — whenever possible — not utilizing natural resources beyond their innate capacity for regeneration or to a point where the vigor, resilience, and stability of ecosystems is compromised. Where nonrenewable resources are to be utilized, they should be consumed judiciously while simultaneously pursuing the development of alternative renewable resources. Finally, since human systems — like every other system — require constant energy inputs and constantly produce waste, it is imperative that the disposal of waste into external environments (whether they are ecosystems, waterways, or the atmosphere) be done in a way that minimizes resultant system disorder, chaos, and instability. To be specific, waste management should be at the very forefront of all human endeavors to support community needs and activities and should be as vigorously pursued as energy and resource development with a commitment toward minimizing the impact of waste disposal and management upon the vigor, resilience, and stability of surrounding environmental systems and domains.

Maintaining ecosystem health must begin with the recognition that environmental health, broadly conceived, is absolutely dependent upon the maintenance of the health and sustainability of human beings, families, neighborhoods, and communities. Moreover, maintaining the health of human communities further entails promoting and maintaining social, political, and economic stability, since instability in any of these areas automatically affects the health status of community inhabitants. This principle of focusing upon human health is a reflection of the degree to which human illness, ignorance, poverty, and conflict have wreaked havoc upon the world's myriad ecosystems and natural communities and is a direct reflection of the values of social ecology. The maintenance of ecosystem health necessitates committing to all of the previously discussed requirements relating to persons, families, homes, neighborhoods, and communities in pursuit of minimizing human community impact upon surrounding ecological households and in the interest of vouchsafing the vigor, resilience, stability, and sustainability of as many of these neighboring ecological households as possible.

Cosmic Ecology

Cosmic Ecology and the Biosphere

Of all the ecological domains considered so far, none represents as great a challenge to articulating a set of basic ecological needs and expectations as does the domain of cosmic ecology. Cosmic ecology is not only the most far-flung and expansive domain, it is also the most ill-defined and mysterious, and it is the ecological domain seemingly most immune from human influence. It is also unique in that it encompasses "the unknown." In fact, cosmic ecology can also be construed as the "ecology of the unknown."

The basic unit of analysis for cosmic ecology — a unit analogous to personal ecology's "self," social ecology's "family," and environmental ecology's "ecosystem" — is the biosphere. The biosphere is the most complex of any of the units of analysis considered so far, consisting of three subdomains (the lithosphere or planet surface, the hydrosphere or the planet's water bodies, and the atmosphere, which is the gaseous membrane surrounding Earth) and encompassing all living creatures (human and nonhuman), species, families, communities, and ecosystems. As such, the biosphere represents the "climactic" ecological domain on the planet. While there may be other units of ecological analysis appropriate to even farther-flung ecological realms, human experience and existence is ultimately realized as interaction occurs between residents of this terrestrial ecosystem and the rest of the cosmos.

Gaia and the Biosphere

Some might associate the biosphere with Gaia theory, as articulated by James Lovelock and others (Lovelock, 2000, 1988; Kirchner, 1991; Margulis and Lovelock, 1976). Lovelock, who first coined the term "Gaia" as a descriptor for the planet, observes:

The temperature, oxidation, state, acidity, and certain aspects of the [Earth's] rocks and waters are kept constant, and . . . this homeostasis is maintained by active feedback processes operated automatically and unconsciously by the biota. Solar energy sustains comfortable conditions for life. The conditions are only constant in the short term and evolve in synchrony with the changing needs of the biota as it evolves. Life and its environment are so closely coupled that evolution concerns Gaia, not the organisms or the environment taken separately.

The planet then circumscribes a living organism, Gaia, a system made up of all the living things and their environment. (Lovelock, 1988, 19, 40)

This vision of Earth as a living entity encompassing all living things and their environments is very much in keeping with Vernadsky's (1926) vision of three distinct yet interrelated hierarchically ecological realms consisting of the geosphere (inanimate matter), the biosphere (biological life), and the noosphere (human thought) — a hierarchy also promulgated by de Chardin (1960). It is also a perspective consistent with my own understanding of the nature and function of the biosphere as the basic unit of analysis and function in cosmic ecology.

More importantly, however, Gaia is a perspective in keeping with our modern understanding of ecosystems as nested networks of self-organizing dissipative processes and structures oriented toward reliable sources of high-quality energy, materials, and information embedded in the immediate and surrounding environments (Hearnshaw, Cullen, and Hughey, 2003, 15; Kay and Regier, 2000). This definition, discussed earlier in terms of ecosystem sustainability, is consistent with a "Gaian" planetary perspective that assumes an evolutionary viewpoint in which planetary biota and their environments interact with one another as complex open systems in which matter, energy and, it would seem, information are transferred and transformed and in which entropy is created and siphoned away from among the planet's ecosystems — ultimately creating chaos in surrounding environments that eventually revisits all systems (Margulis, 1998; Margulis and Sagan, 1997).

Spaceship Earth

Another concept that has come to be closely associated with Gaia theory is "Spaceship Earth." Kenneth Boulding originally introduced the term as a warning to human beings that they exploit the planet at their own peril. According to Boulding: "The earth has become a single spaceship, without unlimited reservoirs of anything, either for extraction or for pollution, and in which, therefore, man must find his place in a cyclical ecological system which is capable of continuous reproduction of material form even though it cannot escape having inputs of energy" (Boulding, 1966, 303). R. Buckminster Fuller (1969) popularized the term to the point where it has become synonymous with planet Earth. Fuller asserts:

> Spaceship Earth is only eight thousand miles in diameter, which is almost a negligible dimension in the great vastness of space. Our nearest star — our energy-supplying mother-ship, the Sun — is ninety-two million miles away, and the nearest star is one hundred thousand times further away. It takes approximately four and one third years for light to get to us from the next nearest energy supply ship star . . . Spaceship Earth was so extraordinarily well invented and designed that to our knowledge humans have been on board it for two million years not even knowing that they were on board a ship. And our spaceship is so superbly designed as to be able to keep life regenerating on board despite the phenomenon, entropy, by which all local physical systems lose energy. So we have to obtain our biological life-regenerating energy from another spaceship — the sun. (Fuller, 1969, 44–45)

Fuller's depiction of Spaceship Earth is reminiscent of Lovelock's concept of Gaia to the extent that both conceive of the planet as a vibrant, nested ecological system that is constantly importing energy, exporting entropy, and changing and evolving to meet the demands of biota and their ecosystems.

Holons

While Earth can be understandably construed as a superorganism, some controversy has arisen regarding the place of humans on such a living planet. Stan Rowe (2001) has likened human beings to organs in this macro-organism. However, Rowe's understanding of the place of humans in the world is one that has generated considerable controversy. For instance, Ken Wilber (1996) has questioned the wisdom of adopting an orientation like Rowe's, perceiving that such an understanding of human roles and worth runs the risk of inviting an eco-fascism in which humans as individuals and perhaps even as a species could be justifiably sacrificed to save the planet as a whole. When humans — or for that matter any entity — are simultaneously considered to be "whole and individual" and a "part or component" of some larger whole, they can be conceived of as "holons." This term, originally introduced by Arthur Koestler (1990) in *The Ghost in the Machine,* applies to every constituent part of a nested ecology. Therefore, when Ken Wilber questions the wisdom of relating to humans as mere components of a larger organism, he does so out of concern that the comparatively independent autonomous nature of human beings — their sense of "agency" or self-direction — will be lost. This concern also applies to any other constituent part of the biosphere or Gaia whose autonomous independence could be subsumed or negated by its simultaneous identity as a part of an even larger supraorganism.

Wilber's concern also applies to any consideration of cosmic ecology in which the biosphere is considered to be the basic unit of analysis and function. As is the case with virtually every other species and ecosystem, the biosphere can also be conceived of as a holon in that it is both an autonomous and independently existing entity and a part — in fact a very small part — of an enormously larger cosmos. The fact that the biosphere is but a part of a much larger whole should not negate the value and autonomy of the Earth as a distinctly unique entity. This same principle holds for every other ecological unit (self, species, family, community, ecosystem) to the degree that each and every individual and unique ecological domain within a nested ecology, and every species subsumed within these systems, is worthy by virtue of their very existence of being recognized and respected as "wholes" that simultaneously function as "parts" of much larger systems and even organisms.

Philosopher Mark Zimmerman captures the issue of acknowledging the autonomy of entities that, as holons, are both "whole" and "part," observing:

> Holons always involve agency-in-communion. Macroscopic structures
> become environments for microscopic ones, and every system is linked
> with its environment by circular processes. The micro, for example, the

organism, is always in relational exchange with the macro, for example, the biome composing the social holon of which the organism is a member. Whereas the organs of an organism (a compound individual holon) are *parts* of it and thus under its general control, the organisms in an ecosystem are *members* of it and not parts of it, because they are not under such strict control, because the complexity of organisms confers on them a relatively high degree of autonomy, and because ecosystem-organism are correlative aspects of the biosphere. The ecosystem is not more "fundamental" than the organisms within it, because organisms and ecosystem mutually influence and constitute one another. (Zimmerman, 2004, 53)

While Zimmerman never uses the terms "nested hierarchies" or "nested ecology," he does a marvelous job of describing the interrelated, nested character of all ecological systems and environments. His description of holons as "agency-in-communion" (sharing their lives and purpose) is particularly useful in that it captures the autonomy of these entities (regardless of their form), and in using the term "agency" he acknowledges that these units are capable of purposefully directing their energies and resources toward means and ends of their own choosing.

The Biosphere, Planetary Metabolism, and Planetary Ecology

The biosphere's subdomains — the lithosphere, hydrosphere, and atmosphere — interact with one another to create a self-regenerating, dissipative system — one described earlier in terms of energy-matter transfer and waste export. Basic biospheric requirements include those conditions and inputs that facilitate self-regeneration and avoid conditions antagonistic to self-renewal. Arguably, the biosphere's basic needs and requirements also include characteristics indicative of ecosystems, namely vigor, organization, resilience, diversity, constancy, stability, and sustainability. More fundamentally, however, the biosphere's most basic needs and functions include photosynthesis — the process by which plants and bacteria utilize solar energy to chemically transform carbon and water into carbohydrates — and primary production, or the rate at which plants and bacteria transform radiant energy from the sun into organic substances (Townsend, Harper, and Begon, 2000; Wessells and Hopson, 1988).

When these basic processes are considered across the entire biosphere as well as from the perspective of Gaia theory (in which Earth is perceived as a living planetary entity), a new paradigm for understanding the needs of the biosphere emerges that is conducive to promoting the planet's health and sustainability — "planetary metabolism." Planetary metabolism is a major research theme promulgated by the Cooperative Institute for Research in Environmental Sciences (CIRES) at the University of Colorado at Boulder. It is also a metaphor explored by George Williams (1996), who uses the term synonymously with "global metabolism" — a concept originating with the French chemist M. J. Dumas (1834). However, whereas Dumas considered global metabolism to be an aggregate measure of global biotic activities, Williams describes it as "the sum of all biological metabolisms on a global [planetary] scale" (Williams, 1996, 104).

By comparison, researchers Shelly Copley and Carol Wessman of CIRES define "planetary metabolism" as "the complex web of biochemical and ecological processes that occur within the biosphere and the interaction of these processes with the lithosphere, atmosphere and hydrosphere" (Copley and Wessman, 2007). These scientists have undertaken an ambitious research agenda designed to further our understanding of how the biosphere's metabolic processes function and what impact humans have upon this global metabolism. They are also seeking to develop a much clearer understanding of the fundamental processes that energize and drive the biosphere, including the effect of biogeochemical cycles and human activities upon water and soil, the relative roles of biogenic and anthropogenic emission sources in affecting air quality, the effects of land management practices on forest and rangeland environments, and the impact of pollution on fresh- and salt-water systems. In so doing, scientists seek to identify those actions needed to safeguard water quality and the health of aquatic ecosystems, and to utilize scientific data for making air and water management decisions.

These are all fundamental issues necessary for understanding the metabolic needs of the planet and responding effectively to them. The focus of this research is indicative of the major challenge facing any thorough understanding of the biosphere's basic needs and fundamental requirements, namely, that we have an enormous amount to learn about how the planet — as a living entity — functions and responds to human actions as well as to natural disturbances from within and beyond the bounds of the planet.

This emerging body of knowledge regarding the basic needs, functions, and processes of the biosphere represents yet another emerging area of scientific inquiry known as "planetary ecology." E. G. Nisbit of the University of London asserts that from the perspective of planetary ecology "the planet as a whole is the ecosystem, not one locality" (Nisbit, 1991, 87). The concept of planetary ecology is also explored by a number of other prominent scientists including Peter J. Mayhew (2006), Duncan Taylor (1994), and Douglas Caldwell, James Brierley, and Corale Brierley (1985), to name but a few.

Of these proponents of a planetary ecology perspective, the thoughts of David Bowman of Charles Darwin University are particularly forceful. Bowman asserts that the focus of modern ecology must become wider than biodiversity alone and instead engage the biosphere itself as the focus of research and intervention, particularly in regard to managing ecosystems and maintaining biogeochemical cycles for energy, matter, and waste transfers (Bowman, 1998). Bowman's comments reflect his overarching concern for maintaining the planet's health. From this vantage point he observes that "given the rapidity of global environmental change, there is an increasing need to manage ecosystems to maintain planetary health. A narrow fixation on biological diversity cannot provide this perspective. Indeed, in the narrowest conception of the term, it is like amassing a huge coin collection in order to understand the functioning of global economies" (Bowman, 1998, 239).

Bowman's concerns are echoed by David Wilkinson in *Fundamental Processes in Ecology* (2006), in which he identifies the seven fundamental processes for planetary sustainability discussed in the previous chapter. Since Wilkinson's work is at the cusp of scientific innovation and discovery, there is little in the way of corroborating evidence to support his assertions. However, research agendas such as those developed by CIRES researchers Shelly Copley and Carol Wessman (2007) will hopefully expand upon Wilkinson's assertions. Ongoing and future studies of the planet as a living biosphere will become increasingly critical to informing the scientific community what the biosphere minimally requires, thereby insuring that the planet's metabolism is maintained and that it persists as a healthy living entity.

Basic Biosphere Needs

A nested ecological perspective is essentially compatible with Gaia theory to the degree that the biosphere is regarded as a living entity reflecting the entirety of the planet's biotic interactions with terrestrial environments. Both assume that the biosphere consists of a network of self-organizing dissipative processes and structures that tap into reliable sources of energy to ultimately produce life. Like Kenneth Boulding and R. Buckmister Fuller, a nested ecology further assumes that Spaceship Earth is capable of continuously producing life and matter, as long as it can maintain access to the sun's energy and can dispel waste. Humans are deemed to be fellow travelers on this spaceship and, like every other creature on the planet, they possess an intrinsic value beyond the particular ecological or systemic function they may assume. Nested ecology also embraces Zimmerman's (2004) contention that all lives possess "agency-in-communion" — implying that every living thing must adapt and fend for itself and its own needs, but does so in a state of interdependence ("in communion") with every other species and community.

Consistent with its compatibility with Gaia theory, nested ecology embraces the concept of "global" or "planetary" metabolism — the complex interactions of biochemical and ecological processes within the biosphere and its subdomains produced by the activities of biota — recognizing that the dissipative ecological systems that characterize the biosphere must ultimately remain healthy and effectively function as a supraorganism capable of maintaining overall system vigor, organization, resilience, diversity, constancy, stability, and sustainability. It further recognizes that the biosphere must at a very minimum be able to sustain photosynthesis and primary production.

However, though it consists of myriad ecosystems, the biosphere does not function at the same level or in the same fashion as lower-order domains in the nested hierarchy. Bowman brilliantly illustrates this point with his analogy regarding mastering global economies by amassing huge coin collections. Nested ecology is fully cognizant of the risk to approaching the needs of a higher-order ecological domain (such as cosmic ecology) as if its functional organization and needs were essentially the same as those of lower-order domains (such as ecosystems, species, or human communities). Consequently, a nested approach reflects the values of

Vladmir Vernadsky, Peter Mayhew, David Bowman, and others, who assert that a wider perspective is demanded to protect the biosphere — one at least as encompassing as the planet itself. In short, nested ecology at the cosmic-ecology level asserts that the appropriate level of analysis and intervention is the biosphere.

Unfortunately, humans are currently ill-equipped to fully engage in an effort to address these planetary needs. The planet as biosphere is exceedingly complex beyond the current capacity of science to significantly appreciate. Even so, important research efforts are under way that seek to address planetary needs, such as improving our appreciation of biogeochemical cycles and human influences on polluting soil, water, and air, or coming to a better appreciation of the relative contributions of biogenic versus anthropogenic sources of emissions upon air quality.

Cosmic Ecology's Fundamental Householding Needs

Despite our current scientific limitations in understanding the biosphere, there are some ecological processes that we can reasonably assert *are* fundamental and must be maintained if life is to indefinitely continue on Earth. More specifically, Wilkinson's seven fundamental functions (Wilkinson, 2006) (energy flow, multiple guilds, tradeoffs, ecological hypercycles, merged organismal and ecological physiology, photosynthesis, and carbon sequestrations) capture the essence of what the planet requires to remain healthy and sustainable, thereby justifying their application to cosmic ecology. What remains is the need to formulate a set of basic human householding functions within the cosmic domain that are responsive to these seven fundamental processes.

Essential Dimensions of Cosmic Ecology

Assuming that humans interact with their environments in a fashion that is sensitive to the fundamental and basic needs, functions, and processes of the biosphere, a functional, healthy, and sustainable approach to cosmic householding minimally requires that humans functionally engage in homemaking and resource provisioning and maintenance in a fashion that vouchsafes the integrity of the planet's basic needs for self-sustenance. Moreover, this cosmic orientation to householding should promote a sense of fully "being at home" on planet Earth in the physical, emotional, and spiritual sense of the term.

If Kenneth Boulding and R. Buckminster Fuller are correct in their analogies, then the biosphere is our home and we — as a species — are in the process of securing the planet as such. Even so, the planet Earth is our home only to the degree that each and every person establishes a specific "sense of place" upon it — a niche from which the rest of the biosphere is conceptualized and experienced, but nevertheless a terrestrial address denoting a specific residence on the planet. As Wendell Berry (1996) has so eloquently observed, citizenship on Earth is not an abstract affair. It occurs in a specific locale and involves specific obligations.

Homemaking on Planet Earth

Since all human beings possess a specific geographic address upon the planet, they are of necessity required to deliberately and instrumentally modify the planet's environment to make it homelike. Fortunately, the geosphere that is the Earth is resilient and resistant enough to tolerate significant perturbation. However, if human disturbances are so significant as to interfere with the energy needs of the biosphere, or so extensive so as to create waste and spent energy beyond the biosphere's capacity to transform or dispel the waste, then fundamental processes critical to the Earth's existence may be thwarted and the vigor and vitality of the planet compromised temporarily or even permanently. Consequently, securing a home on planet Earth requires a new vision of what home entails, and for modern man home is — of necessity — the planet Earth. Homemaking on planet Earth is not a discretionary activity, as it is at every lower ecological level. Barring the capacity to identify other habitable planets or realistically develop the requisite technology for living in space, human beings are planet-bound. Therefore, they are required to adapt their lifestyles to accommodate the inherent ecological limitations that the planet imposes upon them and from which they cannot escape.

Provisioning and Maintaining Human Households on the Planet

All of the provisions needed for human existence on Earth are to be found upon and within the planet. However, these "provisions" occur in the form of natural resources that must be transformed and refined for human use. Moreover, these very same resources are utilized by every other species household and community on the planet. Consequently, overindulging in the utilization of natural resources occurs at the expense of other species and communities and also depletes the planet Earth itself.

Recognizing that the Earth is essentially a supraorganism — a biospheric living entity — and that all life on the planet is a part of this supraorganism, human harm or disruption to any portion of the planet or its life forms constitutes an insult against the biosphere — and this includes the human community. Therefore, while humans have the right and wherewithal to establish, provision, and maintain homes and communities on the face of the planet, their rights are tempered by the equally valid rights of other species households and of the Earth itself. Humans must create and maintain their homes and households within the capacity of the planet and — using the analogy of Michael Zimmerman (2004) — must provision and maintain their homes in "communion" with every other household on Earth.

Vouchsafing Basic Planetary Needs for Self-Sustenance

Recognizing that ecology is a young science and that our corporate ignorance regarding the biosphere substantially outweighs what we currently believe we know, available evidence strongly suggests that basic planetary needs must be identified and satisfied if we are to sustainably continue human habitation upon planet Earth. At minimum, maintaining biospheric health requires that Wilkinson's seven fundamental processes be preserved since they allow the planet's ecosystems to self-organize. While it is unquestionably important to preserve individual species and vouchsafe species diversity, maintaining these vital and fundamental processes is even more important.

Doing so involves recognizing that all human endeavors and communities must become increasingly cognizant of essential planetary processes and — in the words of Wilkinson (2008) — protect "processes" ahead of "entities." Maintaining and sustaining biospheric health also requires humankind to avoid stressing planetary resources beyond the planet's capacity to effectively self-sustain. Efforts in this regard must pay particular attention to tapping fewer nonrenewable natural resources of all kinds, maintaining population and consumption within the carrying capacity of the Earth, and avoiding adding human-induced trauma to the planet regardless of whether such perturbations are the product of industrial or economic activity, war or poverty, carelessness or indifference, or ignorance.

Feeling at Home on Spaceship Earth

The final requirement incumbent upon human communities is the necessity for individual human beings to feel at home on Spaceship Earth. Step one in realizing this outcome is recognizing that home is not just one's domicile, backyard, development, neighborhood, city, state, region, or nation. Especially in an era of global markets, communication, and travel, home is the entire planet. Quite literally, saving any part of the world requires focusing upon the entire planet, since we now realize that all individuals, families, communities, and ecosystems are integrated and interrelated. Similarly, feeling at home on planet Earth requires that we accept the fact that ours is a shared home — a communal abode. If ecology is construed as "householding," then Earth must be considered a household with many rooms and roomers. Being at home means getting over the penchant to approach home as "home alone."

Embracing the Ecology of the Unknown

In concluding this book, I was struck with how much we yet have to learn about ecosystems, the biosphere, the universe, and ourselves. At times when I sat at my computer, searching for words to describe a nested approach to ecology, I found myself frustrated in trying to clarify what our human responsibilities actually involve relative to the planet as a whole and its myriad ecosystems, households, and species. My frustration reflected not merely a dearth of words with which to express myself, but more profoundly a dearth of available data and information regarding how these ecological domains actually function and what they minimally require to continue doing so. I would imagine that the reader who has perused the pages to this chapter has also been impressed with how conceptually diffuse ecological theories, processes, models, and findings become as one progressively moves from the level of self-ecology to social, environmental, and cosmic ecology.

Eventually what we actually know about the world and beyond trickles to a halt, and we are left with the age-old problem of dealing with the complete unknown. Predictably our innate human response to this state of profound ignorance is to experience a sense of "ontic," "ontological," or existential anxiety. Ontological anxiety results from a fear of "nonbeing" — an emotion readily conjured when confronted with the seemingly limitlessness of the universe and our small and comparatively insignificant relationship to it (Tillich, 1957, 1952). "Ontic" anxiety, by comparison, involves a perceived threat to self, particularly in regard to self-preservation, the potential for self-enhancement (Hendrix, 1967), or more simply, the anxiety related to human "fate and death" (Tillich, 1952, 42). Existential anxiety is similar to these other forms and represents a sense of profound uneasiness and dread that is provoked by the vicissitudes and uncertainties of the human condition itself (Kierkegaard, 1944).

Existential, ontic, and ontological anxiety are as old as humans and human cultures and have not been significantly abated by the growth of science, reason and technology, primarily because death, disease, disaster, and deprivation remain as ever present and constant as the mysteriousness of the universe. Such anxiety is particularly pronounced when one stands upon the foundation of the biosphere, gazes outward into the cosmos, and becomes overwhelmed by its seemingly endless breadth (Berger, 1993). Admittedly, staring into the unknown is made measurably bearable by having a cosmological perspective that links the known with the "suspected" and ultimately the unknown and unsuspected. However, there is a point beyond which reason can serve no further purpose in alleviating anxiety, and some other alternative is required.

Of course, an obvious solution is simple denial (Becker, 1973; Freud, 1931) — living as if knowledge is complete and life is fully predictable and without end. For many, denial represents an elegant solution for coping with profound anxiety about living and dying. However, for the thoughtful, fretful, and curious, denial is an insufficient response, so spirituality and religion have arisen to bridge the gap and serve as a balm to humanity's anxious and troubled soul.

In chapter 5 I discussed religion, spirituality, and cosmic ecology in great detail, so I won't reiterate that discussion. Nevertheless, one reality stands out, namely, that beyond the point where science and reasoned theory no longer function, human beings will inevitably utilize conjecture, superstition, belief, and faith in a sincere effort to make sense of their world and their own place within it. Whether spirituality, religion, and faith are actually grounded in the divine, or whether they conveniently serve as a salve for a tortured mind and soul (Freud, 1939) is largely irrelevant to the reality that human beings over the expanse of history have inevitably utilized these tools to cope and make sense of their world. The modern era is no exception to this rule.

Without the coping mechanisms of denial, spirituality, and religion, humanity might have found itself rendered impotent, resigned, and even withdrawn in its interactions with the planet. Had that been the case, curiosity would have certainly been one of the first human qualities to be sacrificed. Fortunately, spirituality and religion appear to play a major role in aiding people to acquire a sense of place and feel at home in the universe despite the fact that they know they and everything they experience will eventually be transformed and die. Science also contributes to a sense of optimism that the unknown is merely that which is not known *yet,* and this optimism serves an ongoing impetus for yet more curiosity and wonder.

As vital as realizing a sense of place and belonging is, spirituality and religion make one additional contribution to dealing with the ecology of the unknown — they contribute to a sense of resilience and stability. When any system exports its entropy beyond its boundaries, it introduces disorder into neighboring systems. This entropy, in turn, contributes to external chaos (disorder) which in time will inevitably be revisited upon the system that jettisoned the entropy in the first place. Consequently, sustainable systems must be highly stable, resistant, and resilient if they are to weather periodic disturbances and upsets. This same process applies to humans, whose anxiety about living is most acute in the face of catastrophic events and onslaughts.

Spirituality and religion provide human beings with a stability and resilience grounded in their perceived connection to and involvement with the eternal forces that create and shape the universe. They also serve to provide a sense of place, purpose, and meaning in the face of fearsome and mysterious life events, thereby enabling people to tolerate and withstand these perturbations. By developing personal capacities for resilience and stability, spirituality and religion serve to strengthen human beings as individuals, families, and communities and engender hope in the face of what could otherwise be experienced as despair.

These basic capacities — resistance and resilience to chaos and disruption, inner stability, and the acquisition of a "sense of place," are integral to living ecologically in the world and essential to transforming households into homes. Without it people inevitably feel out of place, hopeless, and devoid of a future. Such people simply cannot be expected to invest themselves and their values in any person, household, community, or environ. Yet, when people establish a sense of place and belonging, actively and sensitively engage in householding in a locale that feels like home, and ultimately acquire the confidence to believe that they can overcome adversity, they can be expected to relate to their household and those of the surrounding world as an environment in which they belong — one worth caring for and sustaining.

This has been the core assumption underlying and undergirding a nested ecology — namely that people who have a sense of belonging, place, and purpose in the world and who have had the opportunity to develop complete, satisfying, healthy, and sustainable households at the personal and social ecology level will, as a consequence, be motivated and capable of developing and relating their human households with those of other species and communities. Assuming that such nested relationships can be realized and sustained, humans as individuals and as a species should be expected to conclude that Earth in its totality and wholeness is their true abode and come to regard its health and its interests as their own.

Arriving at this conclusion, however, will occur against the backdrop of uncertainty, fear, dread, and anxiety of every form. It will require courage — what Tillich (1952) calls "the courage to be" — in the face of the fearsome and unknown. It will also require curiosity, imagination, perseverance, dedication, and the capacity to regard one's world from micro- to macro-manifestation with constant wonder and undying, unyielding affection. Ultimately, the realization of these and other admirable human traits and ecological values and habits requires that they be nurtured and communicated within the context of a web of mutually supportive and interdependent households, situated like one nested box within another in a seemingly endlessly expansive set of Chinese boxes — a powerful ecological configuration that extends from the self to the stars and beyond.

Epilogue

Pragmatism and Collaboration

This book has attempted to provide a new perspective on human beings and their place on the planet and in the cosmos. My intent has been to impart a pragmatic ecological cosmology that can be intuitively utilized by persons across a wide array of political, philosophical, and spiritual/religious persuasions. What I have assiduously avoided doing, however, is prescribing a particular ecological philosophy or course of action. More specifically, I have avoided asserting any particular ecological ideology, believing instead that thoughtful and ethical people possess the capacity to apply their own values to the problem of protecting the natural resources of the planet.

This is, I believe, the hallmark of pragmatic approaches to solving ecological problems — namely, seeking to achieve consensus and collaboration among people from diverse ideological backgrounds rather than attempting to convert as many as possible to one particular and favored value set. Pragmatists are consensus builders and collaborators, not ideologues or dictators of values. Chris Ansell and Alison Gash, prominent political scientists from the University of California at Berkeley, best describe these pragmatist traits when they observe:

> Pragmatism wants to focus conflict on productive and tractable disputes and to steer antagonists away from dogmatic or institutionalized conflict. Pragmatism suggests that although individuals and groups may have "interests," these interests are typically more ambiguous and

malleable than rational choice perspectives typically allow, and are hence open to reinterpretation through interaction and deliberation with others with different or even opposing perspectives. Pragmatism suggests that more fruitful conflict is made possible through face-to-face social interaction and communication. (Ansell and Gash, 2006, 3)

Ultimately, pragmatism produces collaboration among stakeholders and promotes processes leading to mutually acceptable — not ideal — solutions. This is the application of "satisficing" solutions to ecological problems. Pragmatism assumes a "sanguine" as compared to a "dour" perspective on the capacity of individuals and groups to fruitfully resolve environmental problems in the belief that ultimately "antagonistic stakeholders may be able to transcend intractable disputes and achieve mutual gains through face-to-face deliberation" (Ansell and Gash, 2006, 3).

Pragmatic ecological approaches ultimately produce interventions to remedy ecological problems. Such approaches are collaborative and consensual and never unilateral. They operate upon the assumption that problem-solving skills can be generally acquired and improved upon by most persons as they learn how to systematically analyze and ameliorate troublesome situations by applying provisional remedies. In the world of pragmatics, there is no such thing as an ideal or perfect solution to a problem, only the best possible solution for the present given available time, knowledge, participants, and resources.

Pragmatic interventions depend upon consensus building among a diverse group of stakeholders to ultimately produce an explicit formal agreement suggesting how communities should proceed. This explicit and formal agreement, in turn, serves as a "decision rule" upon which public policy can be developed. Gary Coglianese distinguishes between consensus building and other forms of public participation, observing that when consensus building is utilized

> unanimity becomes the basis for making a policy decision. In this way, consensus-building is different from other forms of public participation because these other forms do not necessarily seek agreement nor dictate unanimity for reaching a decision. The goal of consensus-building, in contrast, is not simply to engage in deliberation and gain public input in order to establish a more informed policy decision. Rather, the goal is

to forge an explicit agreement among all of the participants, with the expectation that this agreement will form the basis for public policy. (Coglianese, 2001, 4–5)

Such explicit agreements must be achieved by bringing people of diverse backgrounds, values, and beliefs together. However, the greater the range of values and beliefs held by the stakeholders, the more difficult it becomes to arrive at an agreement. For this reason, some in the ecology community advocate actively influencing public opinion via the process of environmental education in the interest of creating greater philosophical unanimity.

Environmental Education as Character Reeducation

Mitchell Thomashow's work is exemplary of this approach. Thomashow, in his popular book *Ecological Identity* (1996), presents a variety of approaches to what he calls "ecological identity work" (EIW), a process he developed "to provide the language and context that connect a person's life choices with his or her ecological worldview, serving as a guide that coordinates meaning, a transition to a new way of seeing oneself in the world" (Thomashow, 1996, 6). Thomashow shares a common concern with my own efforts in this text in that both of us are interested in assisting people to begin seeing themselves in a new relationship to the environment. However, while I employ developmental psychology as a vehicle for helping individuals understand themselves relative to their personal ecologies, Thomashow employs psychotherapeutic techniques to facilitate behavioral and, I would argue, character transformation and change.

"Ecological identity work"—which developed to promote PIW or "political identity work" (developing the political will to protect "the commons" and to acquire a sense of "ecological citizenship"), is a term similar to the terminology traditionally utilized within the psychotherapy literature. In this regard it reflects Thomashow's comparably "clinical interventionist" approach. In fact, Thomashow acknowledges that EIW has "therapeutic implications in the sense that people want to heal themselves through their experience of nature, which may summon up painful feelings of loss as well as expressions of joy and happiness" (Thomashow, 1996, 7). EIW is very much a psychotherapeutic approach designed to reeducate and psychologically reorient individuals regarding their relationship to the environment. In this regard, it is essentially "character reeducation," similar to approaches utilized in the management of character disorders (Ramirez, 1968).

The principle tool Thomashow uses in EIW is "introspection" — a process aided by the use of visual and conceptual aids, guided imagery, thought experiments, and other exercises. This process of introspection is what most closely links "ecology identity work" to other psychotherapeutic techniques that also rely upon introspection and reflection. Thomashow acknowledges that introspection associated with EIW can very well "open up a Pandora's box" of emotion and therefore must be undertaken in a fashion that guarantees the emotional safety of the participants and promotes open dialogue and self-expression. These necessary conditions for engaging in EIW are essentially identical to those psychotherapists utilize in conducting their therapeutic work.

In fact, the key word in Thomashow's term "ecological identity work" is the word "work" — a term that has been regularly employed by a wide variety of psychotherapy practitioners. For instance, the term "body work" is used in psychotherapy to refer to the application of "bioenergetics" — as developed by Alexander Lowen (1994) — a process in which the relationships between soma and psyche are unlocked, unleashed, and reintegrated. Likewise, "family work" is a term utilized by family therapists such as Salvador Minuchin (1998), Virginia Satir (1988), Murray Bowen (1994), and others to describe an approach in which an individual's identity, personality, and behavior are conceptualized within the context of the family "system." Comparatively "group work," as conceptualized by Samuel Gladding (2002), is widely used in social-work therapies and seeks to help the individual understand his or her identities, attitudes, and actions against the backdrop of other social groups and organizations. Finally, the term "dream work" is used by psychotherapeutic practitioners influenced by Carl Jung (1978) who seek to orient themselves and their interrelationships to what has been conceived of as the "collective unconsciousness," which some believe actually includes nonhuman and ecological entities. This latter psychotherapeutic school is particularly important because of its influence upon the nascent discipline of eco-psychology. Furthermore, the manner in which it relies upon introspection and reflection are very reminiscent of Thomashow's approach in EIW.

A careful reading of *Ecological Integrity* underscores the degree to which Thomashow consistently utilizes therapeutic language and imagery throughout the narrative (Thomashow, 1996). Thomashow's use of such language is undoubtedly related to the influence of eco-psychology as proffered by Theodore Roszak (2001) and his intellectual followers. Eco-psychology adherents are often referred to as "practitioners"—a term usually reserved for clinicians such as the aforementioned schools of psychotherapy—and typically see themselves as engaged in a form of ecological clinical intervention on behalf of the planet.

By comparison, my own efforts in *Nested Ecology* are more broadly defined and are intended to help people make intellectual connections, clarify their ecological worldviews, and transform their perspectives and values regarding the environment into action. My underlying assumption is that time spent upon these pursuits will ultimately result in the exhibition of more ecologically sustainable and sound actions and policies, thereby preparing people for participation in collaborative and consensus-building initiatives. My approach is not as immediately and directly interventionist as Thomashow's—nor am I seeking to produce an activist or ideologue per se. Instead, I seek to redefine ecological perceptions and relationships as a precursor to enabling people to make thoughtful, systemic, and strategic ecological decisions and to enable them to engage in informed collaboration with others.

Beyond Environmentalism

A more basic difference between Thomashow's approach and my own is that his work is designed to teach "environmentalism" and to create "environmentalists." By comparison, I am disinterested in producing "environmentalists." Instead, I am interested in engaging people from a variety of philosophical perspectives regarding caring for the Earth, regardless of whether they call themselves conservationists, preservationists, environmentalists, Christians, Muslims, Buddhists, or whatever. Admittedly, part of the problem in critiquing Thomashow's work in this regard is that he never clearly defines "environmentalism," even though he liberally uses the term throughout his book. He does goes to great lengths to describe environmentalism using an analogy of a tree with roots, a trunk, branches, leaves, and flowers. In this way he does a very good job of presenting a paradigm for understanding the relationship between older and newer ideas about the environment. He also makes a concerted effort to underscore the diversity of thought to be found within the environmentalist community, observing that (using his tree analogy) "there are many kinds of trees in this forest. Often, radical approaches set up false dichotomies to substantiate their own positions. Deep ecology critiques shallow ecology, assuming that soft reforms reiterate old paradigms. Yet much can be learned from the conservationist tradition of American environmentalism. Concepts such as stewardship and sustainability have great depth, with much room for diverse interpretations and many paths to ecological identity" (Thomashow, 1996, 64). Even so, ecological identity work occurs solely within the framework of "environmentalism," and despite Thomashow's efforts to suggest just how diverse environmentalist thought may be, I resist placing people within that categorization primarily because there are so very many people on the planet who need to be engaged in the process of protecting the Earth's ecosystems but who can in no way be considered "environmentalists."

In fact, many would argue that "environmentalism" is a term that has outlived its usefulness and may in fact be perceived by many in a negative light. Consider, for instance, the report by Michael Shellenberger and Ted Nordhaus, *The Death of Environmentalism* (2005). Like Thomashow, Shellenberger and Nordhaus acknowledge the debt we owe to our intellectual predecessors in the environmental movement, observing: "Those of us who are children of the environmental movement must never forget that we are standing on the shoulders of those who came before us" (Shellenberger and Nordhaus, 2005, 6). In this way they provide an analogy similar to Thomashow's example of a configuration. However, unlike Thomashow, they assert that "at the same time, we believe that the best way to honor their achievements is to acknowledge that modern environmentalism is no longer capable of dealing with the world's most serious ecological crisis" (Shellenberger and Nordhaus, 2005, 6). Having made such a startling statement, they hasten to explain their rationale:

We believe that the environmental movement's foundational concepts, its method for framing legislative proposals, and *its very institutions* are outmoded. Today environmentalism is just another special interest. Evidence for this can be found in its concepts, its proposals and its reasoning. What stands out is how arbitrary environmental leaders are about what gets counted and what doesn't as "environmental." Most of the movement's leading thinkers, funders, and advocates do not question their most basic assumptions about who we are, what we stand for, and what it is that we should be doing. Environmentalism is today more about protecting a supposed "thing" — "the environment" — than advancing the worldview articulated by Sierra Club founder John Muir, who nearly a century ago observed, "When we try to pick out anything by itself, we find it hitched to everything else in the Universe." (Shellenberger and Nordhaus, 2005, 8)

I fully concur with this critique and believe that the concept of environmentalism is delimited not only because it excessively narrows the issue of maintaining the health of the biosphere to "the environment" as a thing, but because the very scope of environmentalism fails to reflect the extent to which the environment as a "thing" is inextricably connected to "everything else in the Universe." My efforts in this book have been about demonstrating just how interconnected the various ecological realms are to one another, and the concept of "environment" or "environmentalism" simply fails to capture either the scope or the nuance of the process we are corporately engaged in. Though well intentioned, Thomashow and others like him who seek to educate and produce "environmentalists" underestimate the scope of the larger ecological endeavor. They are particularly blinded to the reality that the very process of defining themselves and their students as "environmentalists" posits that there is a separate class of persons who must be "nonenvironmentalists" — thereby insuring that a chasm is created between the two groups that resists the very collaboration that pragmatists such as myself desire to engage in.

Once again Shellenberger and Nordhaus clearly perceive the difficulties associated with segregating environmentalists from nonenvironmentalists:

The arrogance here is that environmentalists ask not what we can do for non-environmental constituencies but what non-environmental constituencies can do for environmentalists . . . The environmental movement's incuriosity about the interests of potential allies depends on it never challenging the most basic assumptions about what does and doesn't get counted as "environmental." Because we define environmental problems so narrowly, environmental leaders come up with equally narrow solutions. In the face of perhaps the greatest calamity in modern history [global warming], environmental leaders are sanguine that selling technical solutions like florescent light bulbs, more efficient appliances, and hybrid cars will be sufficient to muster the necessary political strength to overcome the alliance of neoconservative ideologues and industry interests in Washington, D.C." (Shellenberger and Nordhaus, 2005, 9–10)

Given the high stakes involved in our modern ecological crisis, I would have to agree that "environmentalist" and "environmentalism" have become loaded terms that may serve as impediments to achieving a consensus on how to protect and sustain our planetary home.

A Pragmatic Ecologist's Blueprint for Ecological Action and Intervention

Unlike Mitchell Thomashow, I have not sought to write an interventionist text or one dedicated to producing "environmentalists" per se. In fact I have long believed that one of the major omissions in the ecological literature is an emphasis upon an integrated ecological worldview that transcends the simplistic and narrow association of the term "ecology" with "environment." Consequently, I have set out to present a way of looking at oneself in relation to the world and indeed the cosmos. My goals in writing *Nested Ecology* have been more cosmological and philosophical in scope than interventionist, and I have done so in the interest of preparing the way for dialogue, collaboration, and consensus building.

I recognize, however, that in adopting a pragmatic perspective to ecological philosophy and cosmology, I must eventually articulate an interventionist course of action that translates philosophy and cosmology into public policy and practice. The specifics of my own work in this area are forthcoming. However, a preliminary blueprint of what I believe is required of a pragmatic ecologist with a nested ecological worldview would essentially look like the following.

Instead of being grounded in a psychotherapeutic mindset bent upon influencing people to adopt an "environmentalist" philosophy I would prefer to promote ecological virtue among individuals and seek to formulate action plans for ecological protection and restoration through collaborative processes based upon ideological diversity and consensus building. In that regard, I would celebrate the diversity of ecological thought to be found within the larger community rather than attempt to instill a particular ideology — such as "environmentalism" — within individuals.

Practically speaking, such an orientation implies common cause among a wide array of individuals and interest groups and seeks to engender participation from a variety of ideological, spiritual, and religious perspectives. In so doing, I would immediately forgo any notion that ideal solutions to ecological problems are forthcoming, and instead pursue a "satisficing" approach to decision making and policy formulation in the hope that over time better and more complete solutions to ecological problems can be realized. As a corollary to this approach, I would refrain from discounting, castigating, or eliminating people or groups from consideration and collaboration in favor of seeking to find areas upon which agreement and action could be predicated.

Although I would utilize a distinctly psychological approach (rather than a psychotherapeutic one) in conceptualizing the place of the individual within the plethora of nested ecology systems, the overall paradigm I would employ in developing a strategy and plan for ecological action would emanate from systems theory — particularly social system theory. Systems theory is the ideal conceptual approach for integrating human knowledge and experience across every known realm and is the essential level of analysis upon which any successful ecological intervention can occur. Moreover, systems theory is most descriptive of natural ecosystems. Unquestionably any effective ecological action to promote the health and well-being of the biosphere and its inhabitants must be directed toward systems at every level (family systems, social systems, ecosystems, and biospheric systems).

Finally, I would apply consensus-derived solutions regarding ecological problems to public policy considerations while paying particular attention to the socioeconomic ramifications of public policy upon people and nations. I would do so based upon my fervent belief that a necessary prerequisite for sustaining the planet's ecosystems and creatures is to make certain that as many of the planet's human inhabitants as possible achieve a level of affluence that insures that they will not despoil the environment as a result of war or famine and will acquire a level of economic security, education, and freedom requisite for them to fulfill their civic responsibilities as stewards of the Earth's environs. I would extend my concept of "necessary anthropocentrism" to apply to the societies and economies of the world and will assert that affluent citizens are in a better position to consider and act upon the interests of the planet than are poor, uneducated, violence-driven citizens lacking even basic economic resources.

Final Thoughts

At some point I anticipate developing these themes in greater detail. In the interim, however, I would like to leave the reader with this concluding observation from John Muir: "I have never yet happened upon a trace of evidence that seemed to show that any one animal was eager for another as much as it was made for itself . . . No matter, therefore, what may be the note which any creature forms in the song of existence, it is made first for itself, then more and more remotely for all the world and worlds" (Muir, 1918, 7). Muir's words are profound in their simplicity. All creatures are indeed created first for themselves and thereafter "more and more remotely for all the world and worlds." Human beings are no exception to this truism. We are destined to relate to the world first and foremost from the vantage point of our own human anthropocentric interests and only more remotely from the perspective of the creatures and environs surrounding us. It is vital that we come to accept this basic human trait for what it is. However, given the current state of human knowledge, culture, and science during this modern era, we find ourselves living closer to and more intimately dependent upon a much wider range of environs than ever before in our existence. Because of this, it is incumbent upon us as individuals, families, and communities to quickly recognize and acknowledge our creatureliness, to accept our species-derived need to principally care for self and others like us, and thereafter to understand that our future upon the planet is entirely dependent upon our readiness to extend care and consideration for all around us that is not human but upon which our very existence depends.

The world is no longer as remote as it once was, and everything throughout the biosphere is much more immediate to us than we ever imagined before. This recognition should orient us regarding our place in the world and motivate us to move beyond philosophies, politics, economics, religion, and even science to seek consensus among ourselves as the people of planet Earth regarding how we will care and tend for our worldly home. If we fail to adopt this nested perspective of our place on Earth and narrowly follow our creaturely instincts to only care for ourselves and "ours," then we may unwittingly choose to eradicate our species from a world that will continue to thrive, live, and grow long beyond our demise. The choice is ours.

References

....

Chapter 1. Developing a Practical and Sustainable Ecology

Berry, Wendell. 1996. *The unsettling of America.* San Francisco: Sierra Club Books.

Bookchin, Murray. 2001. "What is social ecology?" In *Environmental philosophy,* 3rd edition, ed. Michael Zimmerman, 354–73. New York: Prentice Hall.

——. 1996. *The philosophy of social ecology.* New York: Black Rose Books.

——. 1994. "Will ecology become the dismal science?" In *Environmental ethics and policy book,* ed. Donald VanDeveer and Christine Pierce, 230–34. Belmont, WA: Wadsworth.

——. 1972. *Post-scarcity anarchism.* Berkeley, CA: Ramparts Press.

Boulding, Kenneth E. 1956. "General systems theory — the skeleton of science." *Management Science* 2:197–208.

Bronfenbrenner, Urie. 2005. *Making human beings human: Bioecological perspectives on human development.* Thousand Oaks, CA: Sage Publications.

——. 1978. "Lewinian space and ecological substance." *Journal of Social Issues* 33 (4): 199–212.

Callicott, J. Baird. 1996. "Do deconstructive ecology and sociobiology undermine Leopold's land ethic?" *Environmental Ethics* 18:353–73.

Clark, John. 1990. "What is social ecology?" In *Renewing the Earth: The promise of social ecology,* ed. John Clark, 5–11. London: Green Print.

Clayton, Philip. 2004. "Panentheism in metaphysical and scientific perspective." In *In whom we live and move and have our being,* ed. Philip Clayton and Arthur Peacocke, 73–91. Grand Rapids, MI: William B. Eerdman.

Clayton, Philip, and Peacocke, Arthur, eds. 2004. *In whom we live and move and have our being.* Grand Rapids, MI: William B. Eerdman.

Daly, Herman E. 1997. *Beyond growth: The economics of sustainable development.* Boston: Beacon Press.

Devall, Bill, and George Sessions. 2001. *Deep ecology: Living as if nature mattered.* Layton, UT: Gibbs Smith.

Etzioni, Amitai. 1998. *The new golden rule: Community and morality in a democratic society*. New York: Basic Books.

Haeckel, Ernst. 1866. *General morphology of organisms*. Berlin: Georg Reimer.

Hewlett, Martinez. 2003. "On Polanyi, Clayton, and biology: Some musings of a recovering reductionist." *Tradition and Discovery: The Polanyi Society Periodical* 29 (3): 20–23.

Hormuth, Stefan E. 1990. *The ecology of the self: Relocation and self-concept change*. New York: Cambridge University Press.

Lewin, Kurt L. 1931. "Environmental forces in child behavior and development." In *A Handbook of Child Psychology*, ed. C. Murchison, 94–127. Worcester, MA: Clark University Press.

Littlejohn, Stephen W. 1983. *Theories of human communication*, 2nd edition. Belmont, WA: Wadsworth.

Magee, Patrick T., and Caitlin Walsh. 1998. *Brain dancing*. Bellevue, WA: Braindance.com.

Maslow, Abraham. 1971. *The farther reaches of human nature*. New York: Viking Press.

———. 1968. *Toward a psychology of being*, 3rd edition. London: Wiley Press.

Naess, Arne. 1989. *Ecology, community and lifestyle: Outline of an ecosophy*. Trans. D. Rothenberg. New York: Cambridge University Press.

———. 1973. "The shallow and deep, long-range ecology movement." *Inquiry* 16:97–99.

Parsons, Talcott. 1971. *The system of modern societies*. Englewood Cliffs, CO: Prentice Hall.

Rao, P. K. 2000. *Sustainable development: Economics and policy*. Boston: Blackwell.

Rolston, Holmes, III. 1988. *Environmental ethics: Duties to the values in the natural world*. Philadelphia: Temple University Press.

Scherer, Donald. 1990. *Upstream/downstream: Issues in environmental ethics*. Philadelphia: Temple University Press.

Singer, Peter. 1993. *Practical ethics*. Cambridge: Cambridge University Press.

Stigler, George J. 1946. *The theory of price*. New York: MacMillan.

U.S. Environmental Protection Agency. 2006. "Terms of environment: Glossary, abbreviations and acronyms," www.epa.gov/OCEPATERMS/eterms.html.

Vernadsky, Vladimir. 1926. *The biosphere*. New York: Copernicus, Springer-Verlag.

Von Bertalanffy, Ludwig. 1968. *General systems theory: Foundations, development, applications*. New York: George Braziller.

Zimmerman, Michael E. 1997. *Environmental philosophy: From animal rights to radical ecology*, 2nd edition. New York: Prentice Hall.

Chapter 2. Personal Ecology

Allen, Barbara, and Thomas Schlereth. 1991. *Sense of place: American regional cultures.* Lexington: University Press of Kentucky.

Anton, John P. 1999. "Aristotle and Theophrastus on ecology." In *Philosophy and ecology,* ed. K. Boudouris and K. Kalimtzis, 1:15–27. Athens, Greece: Ionia Publications.

Bennis, Warren G. 1966. "Organizational developments and the fate of bureaucracy." *Industrial Management Review* 7:41–55.

Bennis, Warren G., and Philip E. Slater. 1968. *The temporary society.* New York: Harper Colophon.

Berry, Wendell. 1996. *The unsettling of America.* San Francisco: Sierra Club Books.

———. 1990. *What are people for?* New York: North Point Press.

———. 1984. *Standing by words.* New York: Farrar, Straus, and Giroux.

Biehl, Janet. 1998. *The politics of social ecology.* New York: Black Rose Books.

Bobak, Martin, Hynek Pikhart, Clyde Hertzman, Richard Rose, and Michael Marmot. 1998. "Socioeconomic factors, perceived control, and self-reported health in Russia: A cross-sectional survey." *Social Science and Medicine* 47 (2): 269–79.

Bronfenbrenner, Urie. 2005. *The ecology of human development: Experiments by nature and design.* Cambridge: Harvard University Press.

Cairns, Robert B., Lars R. Bergman, and Jerome Kagan. 1998. *Methods and models for studying the individual.* Thousand Oaks, CA: Sage Publications.

Callicott, J. Baird. 1996. "Do deconstructive ecology and sociobiology undermine Leopold's land ethic?" *Environmental Ethics* 18 (4): 353–73.

———. 1980. "Animal liberation: A triangular affair." *Environmental Ethics* 2 (4): 315–21.

Ceci, Stephen J. 1996. *On intelligence—more or less: A bio-ecological treatise on intellectual development.* Cambridge: Harvard University Press.

Chapple, Christopher Key, and Mary Evelyn Tucker. 2000. *Hinduism and ecology: The intersection of earth, sky, and water.* Religions of the World and Ecology. Cambridge, MA: Harvard University Press.

Cobb, John B. 2002. "Ecology and economy," lecture delivered at Fudan University, Shanghai, June 2, www.religion-online.org/showarticle.asp?title=2221.

Communitarian Network. 1991. *Responsive communitarian platform.* Washington, DC: George Washington University.

Craig, Karrie J., Kelly J. Brown, and Andrew Baum. 2000. *Psychopharmacology: The fourth generation of progress.* Nashville, TN: American College of Neuropsychopharmacology.

Dixon, John, and Kevin Durrheim. 2000. "Displacing place-identity: A discursive approach to locating self and other." *British Journal of Social Psychology* 39:27–44.

Durkheim, Emile. 1951. *Suicide: A study in sociology.* Trans. George Simpson and John A. Spaulding. Glencoe, IL: Free Press.

———. 1947. *The division of labor in society.* Trans. George Simpson. Glencoe, IL: Free Press.

Etzioni, Amitai. 1998. *The new golden rule: Community and morality in a democratic society.* New York: Basic Books.

Haeckel, Ernst. 1866. *General morphology of organisms.* Berlin: Georg Reimer.

Hewlett, Martinez. 2003. "On Polanyi, Clayton, and biology: Some musings of a recovering reductionist." *Tradition and Discovery: The Polanyi Society Periodical* 29 (3): 20–23.

Hormuth, Stefan E. 1990. *The ecology of the self: Relocation and self-concept change.* New York: Cambridge University Press.

Leopold, Aldo. 1949. *A Sand County almanac.* New York: Oxford University Press.

Levine, Norman D. 1975. *Human ecology.* Belmont, CA: Duxbury Press.

Magee, Patrick T., and Caitlin Walsh. 1998. *Brain dancing.* Bellevue, WA: Braindance.com.

Marten, Gerald G. 2001. *Human ecology.* London: Earthscan Publications.

Maslow, Abraham. 1971. *The farther reaches of human nature.* New York: Viking Press.

Mead, George Herbert. 1934. *Mind, self, and society.* Ed. C. W. Morris. Chicago: University of Chicago Press.

Orr, David W. 2002. *The nature of design: Ecology, culture, and human intention.* London: Oxford University Press.

Proshansky, H. M. 1978. "The city and self identity." *Environment and Behavior* 10:147–69.

———. 1976. "Environmental psychology and the real world." *American Psychologist* 31:303–10.

Proshansky, H. M., A. K. Fabian, and R. Kaminoff. 1983. "Place-identity: Physical world socialization of the self." *Journal of Environmental Psychology* 3:57–83.

Rotter, Julian B. 1989. "Internal versus external control of reinforcement: A case history of a variable." *American Psychologist* 45:489–93.

———. 1975. "Some problems and misconceptions related to the construct of internal versus external control of reinforcement." *Journal of Consulting and Clinical Psychology* 43:56–67.

———. 1971. "Locus of control scale." *Psychology Today* (June): 42.

———. 1966. "Generalized expectancies for internal versus external control of reinforcement." *Psychological Monographs* 80:1–28.

———. 1954. *Social learning and clinical psychology.* New York: Prentice Hall.

Slater, Philip. 1992. *A dream deferred: America's discontent and the search for a new democratic ideal.* Boston: Beacon Press.

———. 1990. *Pursuit of loneliness.* Boston: Beacon Press.

Southwick, Charles H. 1996. *Global ecology in human perspective.* London: Oxford University Press.

Stegner, Wallace. 1992. *The sense of place.* New York: Random House.

Steiner, Frederick. 2002. *Human ecology: Following nature's lead.* Washington, DC: Island Press.

Tucker, Mary Evelyn, and Duncan Ruyken Williams. 1998. *Buddhism and ecology: The interconnection of dharma and deeds.* Religions of the World and Ecology. Cambridge: Harvard University Press.

White, Lynn. 1967. "The historical roots of our ecological crisis." *Science* 155: 1203–7.

Chapter 3. Social Ecology

Allen, Barbara, and Thomas Schlereth. 1991. *Sense of place: American regional cultures.* Lexington: University Press of Kentucky.

Allison, James, and Sebastian Moore. 1998. *The joy of being wrong: Original sin through Eastern eyes.* New York: Herder and Herder.

Baugh, Graham. 1990. "The politics of social ecology." In *Renewing the Earth: The promise of social ecology,* ed. John Clark, 97–106. London: Green Print.

Bennis, Warren G., and Philip E. Slater. 1968. *The temporary society.* New York: Harper Colophon.

Berry, Wendell. 1997. *The unsettling of America: Culture and agriculture.* San Francisco: Sierra Club Books.

———. 1996. "Conserving communities." In *Rooted in the land: Essays on community and place,* ed. William Vitek and Wes Jackson, 76–84. New Haven, CT: Yale University Press.

———. 1995. *Another turn of the crank.* New York: Counterpoint Press.

———. 1987. *Home economics.* New York: North Point Press.

Biehl, Janet. 1998. *The politics of social ecology.* New York: Black Rose Books.

Bookchin, Murray. 2003. "Reflections: An overview of the roots of social ecology." *Harbinger: A Journal of Social Ecology* 3 (1): 6–11.

———. 2001. "What is social ecology?" In *Environmental philosophy,* 3rd edition, ed. Michael Zimmerman, 354–73. New York: Prentice Hall.

———. 1997. "Comments on the international social ecology network gathering and the 'deep social ecology' of John Clark." *Democracy and Nature* 3 (3): 154–97.

———. 1996. *The philosophy of social ecology.* New York: Black Rose Books.

———. 1994. "Will ecology become the dismal science?" In *Environmental ethics and policy book,* ed. Donald VanDeveer and Christine Pierce, 230–34. Belmont, WA: Wadsworth.

———. 1993. "What is social ecology?" In *Radical environmentalism: Philosophy and tactics,* ed. Peter C. List. Belmont, WA: Wadsworth.

———. 1982. *The ecology of freedom.* Palo Alto, CA: Ramparts Press.

———. 1980. *Toward an ecological society.* Montreal: Black Rose Books.

———. 1972. *Post-scarcity anarchism.* San Francisco, CA: Ramparts Press.

———. 1970. *Ecology and revolutionary thought.* New York: Times Change Press.

———. 1964. "Ecology and revolutionary thought." *New Directions in Libertarian Thought* (September).

Buber, Martin. 1955. *Paths in utopia.* Boston: Beacon Press.

Callicott, J. Baird. 1996. "Do deconstructive ecology and sociobiology undermine Leopold's land ethic?" *Environmental Ethics* 18:353–73.

Calvin, John. 1559. *Institutes of the Christian religion.* Trans. John Baille, John T. McNeill, and Henry P. Van Dusen. Louisville, KY: John Knox Press.

Chasse, Robert. 1968. *The power of negative thinking, or, Robin Hood rides again.* New York: Situationist International Publications.

Chodorkoff, Daniel. 1990. "Social ecology and community development." In *Renewing the Earth: The promise of social ecology,* ed. John Clark, 69–75. London: Green Print.

Clark, John. 2004a. "The social and the ecological." In *John Clark's social ecology,* http://raforum.info/article.php3?id—article=1044.

———. 2004b. "The ecological self." In *John Clark's social ecology,* http://raforum.info/article.php3?id—article=1047.

———. 1998. "Municipal dreams: A social ecological critique of Bookchin's politics." In *Social ecology after Bookchin,* ed. Andrew Light, 137–91. New York: Guilford.

———. 1990a. "What is social ecology?" In *Renewing the Earth: The promise of social ecology,* ed. John Clark, 5–11. London: Green Print.

———, ed. 1990b. *Renewing the Earth: The promise of social ecology.* London: Green Print.

Edwards, Jonathan. 1821. *A treatise concerning religious affections in three parts.* Philadelphia: James Crissy.

Etzioni, Amitai. 1998. *The new golden rule: Community and morality in a democratic society.* New York: Basic Books.

Friedman, Jeffrey. 1994. "The politics of communitarianism." *Critical Review* 8:297–340.

Hewlett, Martinez. 2003. "On Polanyi, Clayton, and biology: Some musings of a recovering reductionist." *Tradition and Discovery: The Polanyi Society Periodical* 29 (3): 20–23.

Hobbes, Thomas. 1652. *Leviathan,* ed. T. Hobbs and C. B. McPherson (1982). New York: Penguin Books.

Leopold, Aldo. 1966. *A Sand County almanac.* New York: Oxford University Press.

Light, Andrew, ed. 1998. *Social ecology after Bookchin.* New York: Guilford.

Luther, Martin. 1525. *The bondage of the will.* Trans. J. A. Packer and O. R. Johnson (1990). New York: Revel.

MacIntyre, Alaisdar. 1999. *After virtue,* 2nd edition. South Bend, IN: University of Notre Dame Press.

Marten, Gerald G. 2001. *Human ecology.* London: Earthscan Publications.

Mead, George Herbert. 1934. *Mind, self, and society,* ed. C. W. Morris. Chicago: University of Chicago Press.

Orr, David W. 2002. *The nature of design: Ecology, culture, and human intention.* London: Oxford University Press.

Proshansky, H. M., A. K. Fabian, and R. Kaminoff. 1983. "Place-identity: Physical world socialization of the self." *Journal of Environmental Psychology* 3:57–83.

Rousseau, Jean-Jacques. 1762. *'The social contract' and other later political writings.* Trans. Victor Gourevitch (1997). Cambridge: Cambridge University Press.

Russell, Kirk, ed. 1974. *The roots of American order.* La Salle, IL: Open Court.

Schindler, Alfred. 2000. "Zwingli, Huldrych (1484–1531)." In *The Oxford companion to Christian thought,* ed. Adrian Hastings, Alistair Mason, and Hugh Pyper, 765–66. New York: Oxford University Press.

Shantz, Jeff. 2004. "Radical ecology and class struggle: A re-consideration." *Critical Sociology* 30 (3): 691–710.

Simon, Herbert A. 1993. "Satisficing." In *The McGraw-Hill encyclopedia of economics,* 2nd edition, ed. D. Greenwald, 881–86. New York: McGraw-Hill.

Slater, Philip. 1992. *A dream deferred: America's discontent and the search for a new democratic ideal.* Boston: Beacon Press.

———. 1990. *Pursuit of loneliness.* Boston: Beacon Press.

Southwick, Charles H. 1996. *Global ecology in human perspective.* London: Oxford University Press.

Staudenmaier, Peter. 2003. "Economics in a social-ecological society." *Harbinger: A Journal of Social Ecology* 3, no. 1 (Spring): 12–15.

Steiner, Frederick. 2002. *Human ecology: Following nature's lead.* Washington, DC: Island Press.

Stoll, Mark. 2001. "Green versus green: Religions, ethics, and the Bookchin-Foreman dispute." *Environmental history* 6 (July): 412–27.

Suchocki, Marjorie Hewitt. 2001. *God, Christ, Church: A practical guide to process theology.* New York: Crossroad.

———. 1995. *The fall to violence: Original sin in relational theology.* New York: Continuum Press.

Van der Leek, Wilma. 2006. "Life in a machine: The agrarian vision of Wendell Berry." Paper presented at Biblical Faith in the Shadow of Empire, conference meeting of the Institute for Christian Studies (ICS), January 28, Vancouver, Canada.

Watson, David. 1996. *Beyond Bookchin: Preface for a future social ecology.* Detroit: Black and Red.

Zimmerman, Michael E., J. Baird Callicott, George Sessions, Karen J. Warren, and John Clark, eds. 2001. *Environmental philosophy: From animal rights to radical ecology,* 3rd edition. Upper Saddle, NJ: Prentice Hall.

Chapter 4. Environmental Ecology

Aitken, Robert. 1994. *Encouraging words: Zen Buddhist teachings for Western students.* New York: Pantheon Books.

———. 1982. *Taking the path of Zen.* New York: North Point Press.

Berry, Wendell. 2004. *A continuous harmony: Essays cultural and agricultural.* Washington, DC: Shoemaker and Hoard.

Broman, Göran, John Holmberg, and Karl-Henrik Robèrt, 2000. "Simplicity without reduction: Thinking upstream towards the sustainable society." *Interfaces* 30 (3): 13–25.

Canepa, Judy. 1997. "The council of all beings." *Satya* (November): 24–25.

De Silva, Lily. 1987. "The Buddhist attitude toward nature." *Buddhist Perspectives on the Ecocrisis* 343:1–5.

Etzioni, Amitai. 1998. *The new golden rule: Community and morality in a democratic society.* New York: Basic Books.

Freedman, Bill. 1994. *Environmental ecology.* San Diego: Academic Press.

Hardin, Garrett. 1993. *Living within limits: Ecology, economics and population taboos.* New York: New York University Press.

Henning, Daniel H. 2002. *Buddhism and deep ecology.* Bloomington, IN: AuthorHouse.

Hogan, Deirdre. 2007. "Teacher education for sustainable development." Ubuntu

Network CPD Workshop, Irish Aid Department of Foreign Affairs, February 15, www.ubuntu.ie/projects/publications2/final%20draft%20ml.pdf.

Manes, Christopher. 1990. *Green rage: Radical environmentalism and the unmaking of civilization.* Boston: Little, Brown.

Mercier, Jean. 1997. *Downstream and upstream ecologists: The people, organizations, and ideas behind the movement.* Westport, CT: Praeger.

Naess, Arne. 1989. *Ecology, community, and lifestyle: Outline of an ecosophy.* Trans. and ed. D. Rothenberg. New York: Cambridge University Press.

———. 1973. "The shallow and the deep, long-range ecology movement." *Inquiry* 16:97–99.

Nelson, Lance E. 1998. *Purifying the earthly body of God: Religion and ecology in Hindu India.* Albany: State University of New York Press.

Precautionary Principle Project. 2003. "What is the precautionary principle?" A joint initiative of the World Conservation Union, the Wildlife Trade Monitoring Network (TRAFFIC), Fauna and Flora International, Resource Africa, and the European Union, www.pprinciple.net/the—precautionary—principle.html.

Robèrt, Karl-Heinrik, H. Daly, P. Hawken, and J. Holmberg. 1997. "A compass for sustainable development." *International Journal of Sustainable Development and World Ecology* 4, no. 2 (June): 79–92.

Seed, John, Joanna Macy, Pat Fleming, and Arne Naess. 1988. *Thinking like a mountain: Toward a council of all beings.* Gabriola Island, BC: New Society.

Simon, Herbert A. 1993. "Satisficing." In *The McGraw-Hill encyclopedia of economics,* 2nd edition, ed. D. Greenwald, 881–86. New York: McGraw-Hill.

Simon, James P. 2004. *Zen Buddhism and environmental ethics.* Burlington, VT: Ashgate.

Snyder, Gary. 1977. *Mountains and rivers without end.* New York: Counterpoint Press.

———. 1975. *Turtle Island.* Boston: New Directions Press.

———. 1965. *Riprap and Cold Mountain poems.* Washington, DC: Shoemaker and Hoard.

Tucker, Mary Evelyn, and John A. Grim. 1994. *Worldviews and ecology: Religion, philosophy, and the environment.* Maryknoll, NY: Orbis Books.

Wackernagel, Mathis, and William Rees. 1996. *Our ecological footprint: Reducing human impact upon Earth.* Gabriola Island, BC: New Society.

Weaver, Jace. 1996. *Defending Mother Earth: Native American perspectives on environmental justice.* Maryknoll, NY: Orbis Books.

Whitehead, Alfred North. 1979. *Process and reality.* New York: Free Press.

Wimberley, E., and A. Morrow. 1981. "Mulling over muddling through again." *International Journal of Public Administration* 3:483–508.

Chapter 5. Cosmic Ecology and the Ecology of the Unknown

Alpher, Ralph A., and Robert Herman. 1948. "On the relative abundance of the elements." *Physical Review* 74:1737–42.

Barker, E., and M. Matney. 2007. "GEO population estimates using optical survey data." *Orbital Debris Quarterly News* 11 (3): 9–10.

Beisner, E. Calvin. 1997. *Where garden meets wilderness: Evangelical entry into the environmental debate.* Grand Rapids, MI: Eerdmans.

Bekoff, Marc. 2007. *Encyclopedia of human-animal relationships.* Westport, CT: Greenwood Press.

Berry, Thomas. 2000. *The great work: Our way into the future.* New York: Harmony / Bell Tower Press.

——. 1989. *Buddhism.* New York: Columbia University Press.

Bhaskar, Bhagchandra J. 2002. "Ecology and spirituality in the Jain tradition." In *Jainism and ecology,* ed. Christopher Key, 169–80. Cambridge, MA: Harvard University Press.

Bielefeldt, Carl. 2003. "Filling the zen shu." In *Chan Buddhism in ritual context,* ed. Bernard Faure, 179–210. New York: Routledge.

Bloch, Jon P. 1998. "Alternative spirituality and environmentalism." *Review of Religious Research* 40, no. 1 (September): 55–73.

Boeing Aerospace Company. 1982. "Analysis of space systems for the space disposal of nuclear waste follow-on study. Vol. 1, executive summary." *Final report Boeing Aerospace Company: Nuclear and high energy physics category.* Seattle, WA, http://ntrs.nasa.gov/archive/nasa/casi.ntrs.nasa.gov/1982 0014142—1982014142.pdf.

Boff, Leonardo. 1997. *Cry of the Earth, cry of the poor.* New York: Orbis Books.

Bourdeau, P. 2004. "The man-nature relationship and environmental ethics." *Journal of Environmental Radioactivity* 72 (1): 9–15.

Brown, Edward R. 2007. *Our father's world: Mobilizing the church to care for creation.* South Hadley, MA: Doorlight Publications.

Capra, Fritjof. 1979. "Buddhists physics." Schumaker Lecture Series, Bristol University, October 1–8, www.schumacher.org.uk/transcrips/schumlec-79—Bri s—BuddhistPhysics—FCapra.pdf.

——. 1975. *The Tao of physics: An exploration of the parallels between modern physics and Eastern mysticism.* Boston: Shambhala.

Care of Creation. 2007. Madison, WI, http://careofcreation.org/home.cfm.

Carter, Brandon. 1974. "Large number coincidences and the anthropic principle in cosmology." In *Confrontation of cosmological theory with astronomical data,* ed. M. S. Longair, 291–98. Dordrecht, Netherlands: Reidel.

Chapple, Christopher. 2006. "Jainism and ecology: Transformation of tradition."

in *The Oxford handbook of religion and ecology,* ed. Roger S. Gottlieb, 147–59. New York: Oxford University Press.

———. 1992. "Nonviolence to animals in Buddhism and Jainism." In *Inner peace, world peace: Essays on Buddhism and nonviolence,* ed. Kenneth Kraft, 49–62. Albany: State University of New York Press.

COMEST. 2000. "Report of the sub-commission on ethics of outer space." UNESCO Headquarters presentation, Paris, July 10–11, http://unesdoc.unesco.org/imag es/0012/001220/122048e.pdf.

Coppersmith, Jonathan. 2005. "Nuclear waste in space?" *Space Review* (August 22): www.thespacereview.com/article/437/1.

Coward, Harold. 1998. "The ecological implications of karma theory." In *Purifying the earthly body of God,* ed. Lance E. Nelson, 39–60. Albany: State University of New York Press.

Davies, P. C. W. 1992. *The mind of God.* New York: Simon and Schuster.

———. 1982. *The accidental universe.* Cambridge: Cambridge University Press.

Dawkins, Richard. 2006. *The God delusion.* Boston: Houghton Mifflin.

De Chardin, Tielhard. 1961. "Mass of the world." In *Hymn of the universe.* New York: Harper and Row.

Devall, Bill, and George Sessions. 1985. *Deep ecology: Living as if nature mattered.* Salt Lake City, UT: Peregrine Smith.

Dodelson, Scott. 2003. *Modern cosmology.* Burlington, VT: Academic Press.

Drummond, Carl. 2005. "Intelligent design and the future of science education." *Journal of Geoscience Education* 50:134.

Dunlap, Thomas R. 2004. *Faith in nature: Environmentalism as religious quest.* Seattle: University of Washington Press.

Dyson, Freeman. 1979. "Time without end: Physics and biology in an open universe." *Reviews of Modern Physics* 51 (3): 447–60.

Eck, Diane L. 1997. *On common ground: World religions in America.* New York: Columbia University Press.

Eckel, Malcolm David. 1997. "Is there a Buddhist philosophy of nature?" In *Buddhism and ecology: The interconnection of dharma and deeds,* ed. Mary Evelyn Tucker and Duncan Ryuken Williams, 327–49. Cambridge, MA: Harvard University Press.

Evangelical Environmental Network (EEN). 2007. "On the care of creation: An evangelical declaration on the care of creation," www.creationcare.org/resou rces/declaration.php.

Flugel, Peter. 2006. "Jainism and society." *International Quarterly for Asian Studies* 37, no. 1 (May): 198–201.

Forrest, Peter. 1996. *God without the supernatural.* Ithaca, NY: Cornell University Press.

Fox, Matthew. 2000. *Passion for creation: The Earth-honoring spirituality of Meister Eckhart.* Rochester, NY: Inner Traditions.

Fox, Michael Allen. 1999. *Deep vegetarianism.* America in Transition. Philadelphia: Temple University Press.

Geisinger, Alex. 1999. "Sustainable development and the domination of nature: Spreading the seed of the Western ideology of nature." *Boston College Environmental Affairs Law Review* (Fall): 1–25.

Geraci, Robert M. 2006. "Spiritual robots: Religion and our scientific view of the natural world." *Theology and Science* 4(3): 229–46.

Gingrich, Newt. 2007. "Green conservatism: A new way of thinking about the environment." *Human Events* (October 30): www.humanevents.com/article.php?id=23131&page=1#continueA.

Gonzalez, Guillermo, and Jay Richards. 2004. *The privileged planet: How our place in the cosmos is designed for discovery.* Washington, DC: Regnery.

Gosling, David. 2001. *Religion and ecology: In India and Southeast Asia.* New York: Routledge Press.

Grange, Joseph. 1997. *Nature: An environmental cosmology.* Albany: State University of New York Press.

Harris, Sam. 2007. *Letters to a Christian nation: A challenge to the faith of America.* New York: Bantam Press.

Harrison, Edward R. 1981. *Cosmology: The science of the universe.* New York: Cambridge University Press, 1981.

Haught, John F. 1995. *Science and religion: From conflict to conversation.* Mahwah, NJ: Paulist Press.

———. 1994. "Ecology: Restoring our sense of belonging." *Woodstock Report* 38 (June): 3–10.

———. 1993. *The promise of nature: Ecology and the cosmic purpose.* Eugene, OR: Wipf and Stock.

Hawking, Stephen, and Leonard Mlodinow. 2005. *A briefer history of time.* New York: Bantam.

Hawley, John F., and Katherine A. Holcomb. 2003. *Foundations of modern cosmology.* New York: Oxford University Press.

Hetherington, Noriss S. 1993. *Cosmology: Historical, literary, philosophical, religious, and scientific perspective.* Oxford: Routledge Press.

Hitchens, Christopher. 2007a. *God is not great: How religion poisons everything.* Philadephia: Atlantic Books.

———. 2007b. *"God is not great: How religion poisons everything:* Why I wrote this book and why now." *Twelve Books,* www.twelvebooks.com/books/god—not—great.asp?page=behind.

———. 1995. *The missionary position: The ideology of Mother Teresa.* New York: Verso Books.

Hoodbhoy, Pervez A. 2007. "Science and the Islamic world—the quest for rapprochement." *Physics Today* 60, no. 8 (August 3): 49–55.

Ingram, Paul O. 1999. "On the wings of a blue heron." *Crosscurrents* 49:2, www.crosscurrents.org/Ingram.htm.

Islam, Muhammad M. 2004. "Toward a green Earth: An Islamic perspective." *Asian Affairs* 26 (4): 44–89.

Kearns, Laurel. 1996. "Saving the creation: Christian environmentalism in the United States." *Sociology of Religion* 57, no. 1 (Spring): 55–70.

Khalid, Fazlun. 1996. "Guardians of the natural order." *Journal of the United Nations Environmental Programme (UNEP), Our Planet* 8 (2): 8–12.

Knott, Kim. 2000. *Hinduism: A very short introduction.* New York: Oxford University Press.

Kumar, Satish. 2005. *You are, therefore I am: A declaration of dependence.* Devon, UK: Green Books.

———. 1999a. "Reverential ecology," *Resurgence* 195, www.resurgence.org/resurgence/issues/kumar195.htm.

———. 1999b. *Path without destination.* New York: Eagle Brook.

Lancaster, Lewis. 1997. "Buddhism and ecology: Collective cultural perceptions," in *Buddhism and ecology,* ed. Duncan Williams and Mary Evelyn Tucker, 3–18. Cambridge: Harvard University Press.

Lande, Asulv. 2006. "Creation and process theology: A question to Buddhism." In *Buddhism, Christianity and the question of creation: Karmic or divine?* ed. Perry Schmidt-Leukel, 81–92. Aldershot, UK: Ashgate.

Lemaître, Abbé Georges. 1933. "The expanding universe." *Annales de la société scientifique de Bruxelles* A53:51.

Lewis, Martin W. 1994. "Green delusions: An environmentalist critique of radical environmentalism." *Geographical Review* 84, no. 1 (January): 109–11.

Macy, Joanna. 1991. *Mutual causality in Buddhism and general systems theory: The dharma of natural system.* Albany: State University of New York Press.

Margulis, Lynn, et al. 1998. "Foreword to the English translation." In *The biosphere,* by Vladimir Vernadsky. London: Cambridge University Press.

Marsh, N. D., and H. Svensmark. 2000. "Low cloud properties influenced by cosmic rays." *Physical Review Letters* 85:5004–7.

McFague, Sallie. 2000. *Life abundant: Rethinking theology and economy for a planet in peril.* Minneapolis: Augsburg Fortress Press.

McGrath, Allister. 2002. *The reenchantment of nature: The denial of religion and the ecological crisis.* New York: Doubleday.

McGregor, Ronald S. 1997. *The Oxford Hindi-English dictionary.* Oxford: Oxford University Press.

McLaughlin, Andrew. 1993. *Regarding nature: Industrialism and deep ecology.* Albany: State University of New York Press.

Milne, Anthony. 2002. *Sky static: The space debris crisis.* Westport, CT: Praeger.

Minteer, Ben A. 2005. "An appraisal of the critique of anthropocentrism and three lesser known themes in Lynn White's 'Historical roots of our ecologic crisis.' " *Organization and Environment* 18 (2): 163–76.

Moncrief, Lewis W. 1970. "The cultural basis for our environmental crisis: Judeo-Christian tradition is one of many cultural factors contributing to the environmental crisis." *Science* 170, no. 3759 (October 30): 508–12.

Morris, Henry M., and Martin E. Clark. 1976. *The Bible has the answers.* Springdale, AR: Master Books.

Mumme, Patricia. 1998. "Models and images for a vaisnava environmental theology: The potential contribution of srivaisnavism." In *Purifying the earthly body of God,* ed. Lance E. Nelson, 133–61. Albany: State University of New York Press.

Murtazov, A. K. 2003. "Ecology and circumterrestrial space." *Astronomical and Astrophysical Transactions* 22 (4–5): 651–55.

Naess, Arne. 1993. *Ecology, community and lifestyle: Outline of an ecosophy.* New York: Cambridge University Press.

Nanda, Meera. 2005. *The wrongs of the religious right: Reflections on science, secularism, and Hindutva.* Gurgaon, India: Three Essays Collective.

———. 2004. "Dharmic ecology and the neo-pagan international: The dangers of religious environmentalism in India." Paper presented at the 18th European Conference on Modern South Asian Studies, July 8, Lunds University, Sweden, www.sacw.net/DC/CommunalismCollection/ArticlesArchive/072004 —D—Ecology—MeeraNanda.pdf.

———. 2003. *Prophets facing backward: Postmodern critiques of science and Hindu nationalism in India.* Piscataway, NJ: Rutgers University Press.

Narayanan, Vasudha. 2004. *Hinduism.* New York: Oxford University Press.

NASA. 2007. "Orbital Debris mitigation re-emphasized in the new U.S. national space policy." *Orbital Debris Quarterly News* 11 (1): 1.

Nelson, Lance E. 1998. "The dualism of nondualism: Advaita vedanta and the irrelevance of nature." In *Purifying the earthly body of God,* ed. Lance E. Nelson, 61–88. Albany: State University of New York Press.

Ouis, Soumaya Pernilla. 1998. "Global environmental relations: An Islamic perspective." *Newsletter of the Association of Muslim Lawyers,* www.aml.org .uk/journal/4.1/SPO%20-%20Global%20Environment%20Relations.pdf.

Pagels, Heinz. 1987. "A cozy cosmology: The anthropic principle is convenient, but it's not science — reality club lecture." In *Physical cosmology and philosophy*, ed. John Leslie, 174–78. London: MacMillan.

Parel, Anthony J. 2006. *Gandhi's philosophy and the quest for harmony*. Cambridge: Cambridge University Press.

Perry, C. A., and K. J. Hsu. 2000. "Geophysical, archeological, and historical evidence support a solar-output model for climate change." *Proceedings of the National Academy of Sciences* 97:12433–38.

Pew Foundation. 2004. "Religion and the environment: Polls show strong backing for environmental protection across religious groups." *Pew Forum on Religion and Public Life Survey Reports* (October). Washington, DC: Pew Forum on Religion and Public Life.

Plumwood, Val. 1994. *Feminism and the mastery of nature*. Oxford: Routledge.

Pluralism Project. 2007. "Statistics by tradition." The Pluralism Project at Harvard University, www.pluralism.org/resources/statistics/tradition.php# Buddhism.

Polkinghorne, John. 1986. *One world: The interaction of science and theology*. Princeton, NJ: Princeton University Press.

Pompidou, Alain. 2004. "The ethics of outer space" (June). A policy document of the Division of Science and Technology, World Commission on the Ethics of Scientific Knowledge and Technology (COMEST), United Nations Education, Scientific, and Cultural Organizations (UNESCO), Paris, France.

Portee, David S. F., and Joseph P. Loftus. 1993. "Orbital debris and near-Earth environmental management: A chronology." NASA Report RP-1320.

Prime, Ranchor. 1992. *Hinduism and ecology: Seeds of truth*. New York: Cassell.

Reat, Ross N., and Edmund F. Perry. 1991. *A world theology: The central spiritual reality of humankind*. Cambridge: Cambridge University Press.

Robinson, Tri, and Jason Chatraw. 2006. *Saving God's green Earth: Rediscovering the church's responsibility to environmental stewardship*. Norcross, GA: Ampelon.

Rothman, Tony. 1987. "A 'what you see is what you beget' theory." *Discover* (May): 98–99.

Sagan, Carl. 1973. *The cosmic connection: An extraterrestrial perspective*. New York: Cambridge University Press.

Salam, Abdus. 1987. "The future of science in Islamic countries." Paper presented at the Al Islam Islamic Summit, January 27, Kuwait.

Schoch, Richard. 2006. *The secret of happiness: Three thousand years of searching for the good life*. New York: Scribner's.

Sciama, Dennis. 1972. *Modern cosmology*. Lanham, MD: University Press of America.

Scorse, Jason D. 2006. "Religion and environmentalism: A skeptic's view (energy is better spent elsewhere)." *GRIST Environmental News and Commentary,* http://gristmill.grist.org/story/2006/10/15/185447/60.

Scott, Peter. 2003. *A political theology of nature.* Cambridge: Cambridge University Press.

Scudiere, Todd. 2003. *Hindi-English/English-Hindi dictionary and phrasebook.* New York: Hippocrene.

Seed, John. 1994. "Ecopsychology: Psychology in service to the Earth." Paper presented at the Ecopsychology Symposium at the 28th Annual Conference of the Australian Psychological Society, February 10, Gold Coast, Queensland, Australia.

Seielstad, George A. 1983. *Cosmic ecology: The view from the outside in.* Berkeley: University of California Press.

Sessions, George. 1995. *Deep ecology for the 21st century: Readings on the philosophy of the new environmentalism.* Boston: Blackwell.

Shaviv, N. J. and J. Veizer. 2004. "Celestial driver of phanerozoic climate?" *Geological Society of America* 13:4–10.

Sherma, Rita DasGupta. 1998. "Sacred immanence: Reflections of ecofeminism in Hindu tantra." In *Purifying the earthly body of God,* ed. Lance E. Nelson, 89–131. Albany: State University of New York Press.

Shindell, Drew T., David Rind, Nambeth Balachandran, Judith Lean, and Patrick Lonergan. 1999. "Solar cycle variability, ozone, and climate." *Science* 284: 305–8.

Shindell, Drew T., Gavin A. Schmidt, Michael E. Mann, David Rind, and Anne Waple. 2001. "Solar forcing of regional climate change during the maunder minimum." *Science* 294:2149–52.

Silk, Joseph. 1988. *The big bang.* New York: W. H. Freeman.

Sittler, Joseph. 2000. *Evocations of grace: The writings of Joseph Sittler on ecology, theology, and ethics.* Grand Rapids, MI: Eerdmans.

Sivaraksa, Sulak. 2002. "Economic aspects of social and environmental violence from a Buddhist perspective." *Buddhist-Christian Studies* 22:47–60.

Skowlinowski, Henryk. 1990. "For the record: On the origin of eco-philosophy." *Trumpeter* 7:1–11.

Sleeth, Matthew. 2007. *Serve God, save the planet: A Christian call to action.* White River Junction, VT: Chelsea Green.

Smirnov, Nickolay N. 2001. *Space debris.* East Sussex, UK: Taylor and Francis.

Snyder, Gary. 1997. *Mountains and rivers without end.* Berkeley, CA: Counterpoint Press.

——. 1974. *Turtle Island.* New York: New Directions.

Sponberg, Alan. 1994. "Green Buddhism and the hierarchy of compassion." *Western Buddhist Review* 1:131–55.

Stuiver, Martin. 1962. "Variations in radiocarbon concentrations and sunspot activity." *Journal of Geophysical Research* 66:273–76.

Svensmark, Henrik. 1998. "Influence of cosmic rays on Earth's climate." *Physical Review Letters* 81:5027–30.

Svensmark, Henrik, and Nigel Calder. 2007. *The chilling stars: A new theory of climate change.* Cambridge: Icon Books.

Swearer, Donald K. 2006. "An assessment of Buddhist eco-philosophy." *Harvard Theological Review* 99 (2): 123–37.

Tatiia, Nathmal. 2002. "The Jain worldview and ecology." In *Jainism and ecology,* ed. Christopher Key Chapple, 3–18. Cambridge: Harvard University Press.

———, trans. 1994. *That which is: Tattvartha sutra.* Lanham, MD: Alta Mira Press.

Tetreault, Steve. 2006. "Bigger repository backed in study preview: Yucca Mountain could be redesigned to hold up to 628,000 tons." *Las Vegas Review Journal* (April 20): www.reviewjournal.com/lvrj—home/2006/Apr-20-Thu-2006/news/6955820.html.

Tobias, Michael. 2000. *Life force: The world of Jainism.* Fremont, CA: Jain Publishing.

Tucker, Mary Evelyn. 1998. "The philosophy of ch'i as an ecological cosmology." In *Confucianism and ecology,* ed. Mary Evelyn Tucker and John Berthrong, 187–210. Cambridge, MA: Harvard University Press.

Turner, Frederick Jackson. 1920. *The frontier in American history.* Cambridge, MA: Harvard University Press.

U.S. Department of Energy. 2003. "Managing nuclear waste: Options considered." Office of Civilian Radioactive Waste Management, Yucca Mountain Project, Las Vegas, NV, www.ocrwm.doe.gov/factsheets/doeymp0017.shtml.

Verhoeven, Martin J. 2001. "Buddhism and science: Probing the boundaries of faith and reason." *Religion East and West* 1:77–97.

Vernadsky, Vladimir. 1991. *Scientific thought as a planetary phenomenon.* Moscow: Institute for System Analysis of Russian Academy of Sciences.

———. 1926. *The biosphere.* New York: Copernicus, Springer-Verlag.

Viikari, L. 2002. "Environmental impact assessment and space activities." Proceedings of the 34th Cospar Scientific Assembly, the Second World Space Congress, October 10–19, Houston, TX.

Warren, Henry C. 2005. *Buddhism in translations: Passages selected from the Buddhist sacred books.* Whitefish, MT: Kessinger.

Weber, Thomas. 1999. "Gandhi, deep ecology, peace research and Buddhist economics." *Journal of Peace Research* 36 (3): 349–61.

Weinberg, Steven. 1977. *The first three minutes.* New York: Basic Books.

Whitcomb, John, and Henry M. Morris. 1989. *The Genesis flood.* Philipsburg, NJ: P&R.

White, Lynn. 1967. "The historical roots of our ecological crisis." *Science* 155: 1203–7.

Whitehead, Alfred North. 1938. *Modes of thought*. New York: Macmillan.

Williams, Duncan R. 1997. "Animal liberation, death and the state: Rites to release animals in medieval Japan." In *Buddhism and ecology,* ed. Duncan Williams and Mary Evelyn Tucker, 149–64. Cambridge, MA: Harvard University Press.

Williams, William A. 1928. *The Evolution of Man Scientifically Disproved in 50 Arguments,* www.creationism.org/books/WilliamsEvolManDisprvd/8evds10h.htm#intro.

Wilson, E. O. 2007. *The creation: An appeal to save life on Earth.* New York: W. W. Norton.

Yamamoto, Shuichi. 2001. "Mahayana Buddhism and environmental ethics: From the perspective of the consciousness-only doctrine." *Journal of Oriental Studies* 11:167–80.

Chapter 6. Essential Characteristics of Nested Ecology

Allen, Timothy F. H. 1996. *Hierarchy theory: A vision, vocabulary, and epistemology.* New York: Columbia University Press.

Allen, Timothy F. H., and T. B. Starr. 1982. *Hierarchy: Perspectives for ecological complexity.* Chicago: University of Chicago Press.

Antonovsky, Aaron, and Talma Sourani. 1988. "Family sense of coherence and family adaptation." *Journal of Marriage and the Family* 50 (1): 79–92.

Bailey, Robert G. 2002. *Ecoregion-based design for sustainability.* New York: Springer.

Baldwin, Michelle, and Virginia Satir. 1987. *The use of self in therapy.* Philadelphia: Haworth Press.

Bielby, D. D. 1992. "Commitment to work and family." *Annual Review of Sociology* 18:281–302.

Bjonnes, Roar. 2003. "A new vision of development." *Prout Journal* 10 (2): 1.

Bookchin, Murray. 2003. "An overview of the roots of social ecology." *Harbinger: A Journal of Social Ecology* 3 (1): 6–11.

——. 1986. *The modern crisis.* Gabriola Island, BC: New Society.

Bowman, David. 1998. "The death of biodiversity—the urgent need for global ecology." *Global Ecology and Biogeography Letters* 7 (4): 237–40.

Brundtland Commission. 1987. *Our common future: World commission on environment and development (the Bruntland commission).* Oxford: Oxford University Press.

Bull, Hedley. 1977. *The anarchical society.* New York: Columbia University Press.

Campbell, Donald T. 1974. " 'Downward causation' in hierarchically organized biological systems." In *Studies in the Philosophy of Biology*, ed. F. J. Ayala and T. Dobzhansky, 179–86. Berkeley: University of California Press.

Copley, Shelly, and Carol Wessman. 2007. "Planetary metabolism." Cooperative Institute for Research in Environmental Sciences (CIRES), University of Colorado, Boulder, Institute Research Theme on Planetary Metabolism, http://cires.colorado.edu/science/themes/planet/.

Costanza, R., ed. 1992. *Ecosystem health: New goals for environmental management*. Washington, DC: Island Press.

Costanza, R., and M. Mageau. 2000. "What is a healthy ecosystem?" *Aquatic Ecology* 33:105–15.

Darwin, Charles R. 1882. *On the origin of species by means of natural selection, or the preservation of favoured races in the struggle of life*. London: John Murray.

DeFrain, John. 2000. "Creating a strong family commitment: The family comes first." *Nebfact*, University of Nebraska–Lincoln Agricultural Extension Service Bulletin, NF441 (September): www.ianrpubs.unl.edu/epublic/live/g1839/build/g1839.pdf.

Downs, Kimberly J. M. 2004. "Family commitment, role perceptions, social support, and mutual children in remarriage: A test of uncertainty reduction theory." *Journal of Divorce and Remarriage* 40 (1–2): 35–53.

Duhl, Leonard. 2003. *The social entrepreneurship of change*. London: Cogent.

Duhl, Leonard, and Hancock, Trevor. 1997. "Healthy cities, healthy children." In *Industrialized country commentary: The progress of nations 1997*, 59–60. New York: United Nations (UNICEF).

Durand, Rodolphe, and Eero Vaara. 2006. "A true competitive advantage? Reflections on different epistemological approaches to strategy research." *Les cahiers de recherche* (Paper 838), HEC School of Management, Jouy-en-Josas cedex, France, www.hec.fr/hec/fr/professeurs—recherche/upload/cahiers/CR838.pdf.

Dutu, Mircea. 2004. "Constitutionalizing the right to a healthy environment and its implications in Romanian legislation." White paper (January 13), Rural Rehabilitation and Development Ministry (RRDM), Bucharest, Romania.

Feibleman, James K. 1954. "Theory of integrative levels." *British Journal for the Philosophy of Science* 5:59–66.

Fromm, E. 1956. *The sane society*. London: Routledge and Kegan.

Gordon, Larry. 1998. "Environmental health definition." In *An ensemble of definitions of environmental health*, ed. Risk Communication and Education Subcommittee, Environmental Health Policy Committee, U.S. Department of Health and Human Services (November 20): http://web.health.gov/ environment/DefinitionsofEnvHealth/ehdef2.htm.

Gottlieb, Roger S. 2003. *This sacred earth: Religion, nature, environment.* New York: Routledge.

Grizzle, Raymond E. 1994. "Environmentalism should include human ecological needs." *Bioscience* 44 (4): 263–68.

Gunderson, Lance H., and C. S. Holling, eds. 2002. *Panarchy: Understanding transformations in human and natural systems.* Washington, DC: Island Press.

Gunderson, Lance H., C. S. Holling, and S. S. Light. 1995. *Barriers and bridges to the renewal of ecosystems and institutions.* New York: Columbia University Press.

Haley, Jay. 2003. *The art of strategic therapy.* New York: Routledge.

Hammond, H., T. Bradley, J. Kubian, and S. Hammond. 1996. *An ecosystem-based landscape plan for the Slocan River watershed.* Slocan Park, BC: Silva Forest Foundation.

Hawley, Dale R., and Laura DeHaan. 1996. "Toward a definition of family resilience: Integrating life-span and family perspectives." *Family Process* 35 (3): 283–98.

Jolley, John L. 1973. *The fabric of knowledge: A study of the relations between ideas.* London: Duckworth Press.

King, Martin Luther, Jr. 1958. "Advice for living." *Ebony* (January): 34.

Kinzig, Ann, L. Gunderson, A. Quinlan, and B. Walker. 2007. *Assessing and managing resilience in social-ecological systems: A practitioners workbook, version 1.0.* Waterloo, ON: Resilience Alliance.

Klijn, Frans, and Helias A. Udo de Haes.1994. "A hierarchical approach to ecosystems and its implications for ecological land classification." *Landscape Ecology* 9 (2): 89–104.

Kropotkin, Peter. 1902. *Mutual aid: A factor of evolution.* Boston: Porter Sargent.

Langschwert, G. 1998. "Workshop on encouraging local initiatives towards sustainable consumption patterns," Economic Commission for Europe (ECE), United Nations, February 2–4, Vienna, Austria.

Light, Andrew, and Eric Katz. 1996. "Introduction: Environmental pragmatism and environmental ethics as contested terrain." In *Environmental Pragmatism,* ed. Andrew Light and Eric Katz, 1–20. New York: Routledge, Taylor, and Francis.

Linnaeus, Carolus. 1735. *Systema naturae.* Stockholm: Lipsiae.

Lyons, R., and L. Langille. 2000. *Healthy lifestyle: Strengthening the effectiveness of lifestyle approaches to improve health.* Ottawa: Health Canada, Population and Public Health Branch.

Margolis, Joseph. 2003. "Pragmatism's advantage." *Ars disputandi* 3, www.arsdisputandi.org/publish/articles/000126/index.html.

Maslow, Abraham H. 1943. "A theory of human motivation." *Psychological Review* 50:370–96.

McCubbin, Hamilton I., and Joan M. Patterson. 1983. "The family stress model: The double abcx model of adjustment and adaptation." In *Social stress and the family: Advances and developments in family stress theory and research,* ed. Hamilton McCubbin, Marvin Summan, and Joan Patterson, 7–37. Binghamton, NY: Haworth Press.

McGoldrick, Monica. 1998. *Re-visioning family therapy: Race, culture, and gender in clinical practice.* New York: Guilford.

Minuchin, Salvador. 1998. *Family healing: Strategies for hope and understanding.* Cambridge, MA: Harvard University Press.

———. 1986. *Family kaleidoscope.* Cambridge, MA: Harvard University Press.

———. 1974. *Families and family therapy.* Cambridge, MA: Harvard University Press.

Minuchin, Salvador, and H. Charles Fishman. 1981. *Family therapy techniques.* Cambridge, MA: Harvard University Press.

Minuchin, Salvador, Michael P. Nichols, and Wai Yung Lee. 2006. *Assessing families and couples: From symptom to system.* Upper Saddle River, NJ: Allyn and Bacon.

National Environmental Health Association (NEHA). 1996. "Environmental health definition," www.neha.org/position—papers/def—env—health.html.

Norton, Bryan. 2007. "Ethics and sustainable development: An adaptive approach to environmental choice." In *Handbook of sustainable development,* ed. Giles Atkinson, Simon Dietz, and Eric Neumayer. Cheltenham, UK: Edward Elgar.

———. 2006. "The re-birth of environmentalism as pragmatic, adaptive management." University of Washington Oceans to Stars Lecture Series: The Human Imprint, presented by the University of Wisconsin Alumni Association and the University of Wisconsin Earth Initiative, May 10, Seattle, WA, www.law .virginia.edu/pdf/envlaw05conf/norton—bryan.pdf.

———. 2005. *Sustainability: A philosophy of adaptive ecosystem management.* Chicago: University of Chicago Press.

———. 2003. *Searching for sustainability: Interdisciplinary essays in the philosophy of conservation biology.* New York: Cambridge University Press.

Olsen, Jan A., Jeff Richardson, and Paul Menzel. 1998. "The moral relevance of personal characteristics in setting health care priorities." *Social Science and Medicine* 57 (7): 1163–72.

O'Neill, Robert V., Donald L. DeAngelis, Jack B. Waide, and Timothy H. F. Allen, eds. 1986. *A hierarchical concept of ecosystems.* Princeton, NJ: Princeton University Press.

Parker, Kelly A. 1996. "Pragmatism and environmental thought." In *Environmental pragmatism,* ed. Andrew Light and Eric Katz, 21–37. New York: Routledge.

Pattee, Howard H. 1973. "The physical basis and origin of hierarchical control." In *Hierarchy theory: The challenge of complex systems,* ed. H. H. Pattee. New York: George Braziller.

Phelps, Patricia. 2000. "Creating a commitment to family involvement." *Rural Educator* 21 (3): 35–38.

Polanyi, M. 1968. "Life's irreducible structure." *Science* 160:1308–12.

Pope, A. M., M. A. Snyder, and L. H. Mood, eds. 1995. *Nursing, Health, and the Environment.* Washington, DC: Institute of Medicine.

Prezioso, F. A. 1987. "Spirituality in the recovery process." *Journal of Substance Abuse Treatment* 4:233–38.

Rapport, David J. 1998. "Defining ecosystem health." In *Ecosystem health,* ed. R. Rapport, R. Costanza, P. Epstein, C. Gaudet, and R. Levins, 18–33. Malden, MA: Blackwell.

Rapport, David J., Bill L. Lasley, Dennis E. Rolston, N. O. Nielsen, Calvin O. Qualset, and Ardeshir B. Damania, eds. 2002. *Managing for healthy ecosystems.* Boca Raton, FL: CRC Press.

Reeves, Matthew J., and Ann P. Rafferty. 2005. "Healthy lifestyle characteristics among adults in the United States, 2000." *Archives of Internal Medicine* 165:854–56.

Robbins, Michelle. 1996. "Thinking sustainably — sustainable ecosystems." *American Forester* 102, no. 2 (Spring): 33, http://findarticles.com/p/articles/mi—m1016/is—n2—v102/ai—18333986.

Robinson, Lance, Tony Fuller, and David Waltner-Toews. 2003. *Ecosystem health and sustainable livelihoods approaches — a synthesis of the latest thinking dealing with complexity in rural development and agriculture.* Ottawa: Canadian International Development Agency.

Robinson, Nicholas. 1992. *Environmental law lexicon.* Washington, DC: Law Journal Press.

Satir, Virginia. 1988. *The new peoplemaking.* Palo Alto, CA: Science and Behavior Books.

Satir, Virginia, and Michelle Baldwin. 1984. *Satir step by step: A guide to creating change in families.* Palo Alto, CA: Science and Behavior Books.

Schein, Martin W. 1975. *Social hierarchy and dominance.* Hoboken, NJ: Wiley.

Seedhouse, David. 2001. *Health: The foundations for achievement,* 2nd edition. Hoboken, NJ: Wiley.

Sheridan, Susan M., John W. Eagle, and Shannon E. Dowd. 2005. "Families as context for children's adaptation." In *Handbook of resilience in children*, ed. Sam Goldstein and Robert Brook, 165–80. New York: Springer.

Simon, Herbert A. 1973. "The organization of complex systems." In *Hierarchy theory*, ed. Howard H. Pattee, 3–27. New York: George Braziller.

——. 1969. *The sciences of the artificial*. Cambridge, MA: MIT Press.

Spretnak, Charlene. 1986. *The spiritual dimension of green politics*. Rochester, VT: Bear.

Svoboda, Melanie. 2005. *Traits of a healthy spirituality (inspirational reading for every Catholic)*. New London, CT: Twenty-Third Publications.

Theobald, Douglas L. 2004. "29+ evidences for macroevolution: The scientific case for common descent," www.talkorigins.org/faqs/comdesc/.

U.S. Department of Health and Human Services (USDHHS). 2007. *Healthy People 2010*. Washington, DC: Office of Disease Prevention and Health Promotion.

——. 1998. "An ensemble of definitions of environmental health," U.S. Department of Health and Human Services Environmental Health Policy Committee Risk Communication and Education Subcommittee (November 20): www.health.gov/environment/DefinitionsofEnvHealth/ehdef2.htm.

Vaughn, Clark F. 1991. "Spiritual issues in psychotherapy." *Journal of Transpersonal Psychology* 23 (2): 105–19.

Von Bertalanffy, Ludwig. 1968. *General systems theory*. New York: George Braziller.

Wallace, Mark I. 2005. "The earthen spirit: How green spirituality can inform the environmental movement." *Swarthmore College Bulletin* (September 25): 17–23.

——. 2003. "Environmental justice, neopreservationism, and sustainable spirituality." in *This sacred Earth: Religion, nature, environment*, ed. Roger S. Gottlieb, 596–612. New York: Routledge.

Weber, Max. 1947. *Economy and society,* trans. Talcott Parsons. New York: Oxford University Press.

Weiss, Paul A. 1969. "The living system: Determinism stratified." In *Beyond reductionism*, ed. A. Koestler and J. R. Smythie, 3–55. London: Hutchinson Press.

Whitchurch, G., and L. Constantine. 1993. "Systems theory." In *Sourcebook of family theories and methods: A contextual approach*, ed. P. Boss, W. Doherty, R. LaRossa, W. Schumm, and S. Steinmetz, 325–52. New York: Plenum Press.

Wilkinson, David M. 2006. *Fundamental processes in ecology: An Earth systems approach*. New York: Oxford University Press.

Wilson, David S. 1997. "Introduction: Multilevel selection theory comes of age." *American Naturalist* 150 (July): S1–S4.

Chapter 7. The Fundamentals of Nested Ecological Householding

Adams, Bruce. 1995. "Building healthy communities." *Pew partnership for civic change leadership collaboration series* (Winter): 2–17.

Adger, W. Neil. 2000. "Social and ecological resilience: Are they related?" *Progress in Human Geography* 24 (3): 347–64.

Anderson, Greg. 2005. *Thermodynamics of natural systems.* Cambridge: Cambridge University Press.

Anderson, Laurie M., Carolynne Shinn, and Joseph St. Charles. 2002. "Community interventions to promote healthy social environments: Early childhood development and family housing: A report on recommendations of the task force on community preventive services." *Mortality and Morbidity Weekly Report* 51, RR01 (February 1): 1–8.

Arrow, Kenneth B., et al. 1995. "Economic growth, carrying capacity, and the environment." *Science* 268 (April 28): 520–21.

Bakkes, J. A. 1994. *An overview of environmental indicators.* Bilthoven, Netherlands: Rivm Press.

Baumeister, Roy F., and Mark R. Leary. 1995. "The need to belong: Desire for interpersonal attachments as a fundamental human motivation." *Psychological Bulletin* 117:497–529.

Beavers, W. R. 1982. "Healthy, midrange, and severely dysfunctional families." In *Normal Family Processes,* ed. F. Walsh, 45–66. New York: Guilford.

Becker, Ernest. 1973. *The denial of death.* New York: Free Press.

Becker, Gary. 2005. *A treatise on the family.* Cambridge, MA: Harvard University Press.

Begon, M., J. L. Harper, and C. R. Townsend. 2000. *Ecology: Individuals, populations, and communities,* 3rd edition. Sunderland, MA: Sinauer Associates.

Belovsky, Gary E. 2004. "Ecological stability: Reality, misconceptions, and implications for risk assessment." *Human and Ecological Risk Assessment* 8, no. 1 (January–March): 99–108.

Berger, Peter. 1993. *A far glory.* New York: Anchor Press.

Berkowitz, W., and S. Cashman. 2000. "Building healthy communities: Lessons and challenges." *Community* 3 (2): 1–7.

Berry, Wendell. 1996. *The unsettling of America.* San Francisco: Sierra Club Books.

———. 1987. *Home economics.* New York: North Point Press.

Bookchin, Murray. 2003. "An overview of the roots of social ecology." *Harbinger: A Journal of Social Ecology* 3 (1): 6–11.

———. 1986. *The modern crisis*. Gabriola Island, BC: New Society.

Boulding, Kenneth. 1963. "The economics of the coming spaceship Earth." In *Environmental quality in a growing economy*, ed. H. Jarrett, 3–14. Boston: MIT Press.

Bowman, David. 1998. "The death of biodiversity—the urgent need for global ecology." *Global Ecology and Biogeography Letters* 7 (4): 237–40.

Caldwell, Douglas E., James A. Brierley, and Corale L. Brierley. 1985. *Planetary ecology*. New York: Van Nostrand Reinhold.

Canada International Development Agency (CIDA). 1997. "Policy on meeting basic human needs." *Basic human needs—a few facts*. Ottawa: Government of Canada.

Carpenter, S. R. 2003. *Regime shifts in lake ecosystems: Pattern and variation*. Excellence in Ecology Series 15. Oldendorf/Luhe, Germany: Ecology Institute.

Carpenter, S. R., B. H. Walker, J. M. Anderies, and N. Abel. 2001. "From metaphor to measurement: Resilience of what to what?" *Ecosystems* 4:765–81.

Common, M. 1995. *Sustainability and policy: Limits to economics*. Cambridge: Cambridge University Press.

Connard, Christie, and Rebecca Novick. 1996. "The ecology of the family, a background paper for a family-centered approach to education and social service delivery" (February). N.W. Regional Education Laboratory, Portland, OR, www.nwrel.org/cfc/publications/ecology2.html#Key%20Concepts.

Connell, Des W., Paul Lam, Bruce Richardson, and Rudolph Wu. 1999. *Introduction to ecotoxicology*. Hoboken, NJ: Wiley-Blackwell.

Cooley, William Forbes. 2004. *The individual: A metaphysical inquiry*. Whitefish, MT: Kessinger.

Copley, Shelly, and Carol Wessman. 2007. "Planetary metabolism." Cooperative Institute for Research in Environmental Sciences (CIRES), University of Colorado, Boulder, Institute Research Theme on Planetary Metabolism, http://cires.colorado.edu/science/themes/planet/.

Costanza, Robert. 1992. "Toward an operational definition of ecosystem health." In *Ecosystem health: New goals for environmental management*, ed. Robert Costanza and Bryan G. Norton, 236–53. Washington, DC: Island Press.

Costanza, R., and M. Mageau. 2000. "What is a healthy ecosystem?" *Aquatic Ecology* 33:105–15.

Covey, Stephen R., and Sandra M. Covey. 1998. *The seven habits of highly effective families*. New York: St. Martin's Press.

Crowley, Philip H. 1977 "Spatially distributed stochasticity and the constancy of ecosystems." *Bulletin of Mathematical Biology* 39 (2) 157–66.

Curran, D. 1983. *Traits of a healthy family*. Minneapolis: Winston Press.

De Chardin, Telhard. 1960. *The divine milieu*. Mahwah, NJ: Paulist Press.

Duhl, Leonard, and Trevor Hancock. 1997. "Healthy cities, healthy children." In *Industrialized country commentary: The progress of nations 1997*, 59–60. New York: United Nations (UNICEF).

Dumas, M. J. 1834. "Organic chemistry research." *Annals of Physical Chemistry* 56:115–20.

Epstein, N. B., D. Bishop, C. Ryan, I. W. Miller, and G. I. Keitner. 1993. "The McMaster model view of healthy family functioning." In *Normal family processes*, ed. Froma Walsh, 138–60. New York: Guilford.

Etzioni, Amitai. 1998. *The new golden rule: Community and morality in a democratic society.* New York: Basic Books.

Fermi, Enrico. 1956. *Thermodynamics.* New York: Dover.

Fine, M. 1992. "A systems-ecological perspective on home-school intervention." In *The handbook of family-school interventions: A systems perspective*, ed. M. Fine and C. Carlson, 1–17. Boston: Allyn and Bacon.

Flower, Joe. 1993. "Building healthy cities: Excerpts from a conversation with Leonard J. Duhl, M.D." *Healthcare Forum Journal* 36 (3): 48–54.

Freud, Sigmond. 1939. *Moses and monotheism: Three essays.* In *The standard edition of the complete psychological works of Sigmund Freud*, trans. James Strachey, 36–53. London: Hogarth Press.

———. 1931. *Civilization and its discontents.* London: Hogarth Press.

Fuller, R. Buckminster. 1969. *Operating manual for Spaceship Earth.* Carbondale: Southern Illinois University Press.

Gigon, Andreas. 1983. "Typology and principles of ecological stability and instability." *Mountain Research and Development* 3 (2): 95–102.

Gunderson, Lance H., and C. S. Holling, eds. 2002. *Panarchy: Understanding transformations in human and natural systems.* Washington, DC: Island Press.

Gunderson, Lance H., C. S. Holling, and S. S. Light. 1995. *Barriers and bridges to the renewal of ecosystems and institutions.* New York: Columbia University Press.

Gunderson, Lance H., C. S. Holling, L. Pritchard, and G. D. Peterson. 1997. "Resilience in ecosystems, institutions and societies." Discussion paper 95. Stockholm: Beijer International Institute of Ecological Economics.

Hammond, H., T. Bradley, J. Kubian, and S. Hammond. 1996. *An ecosystem-based landscape plan for the Slocan River watershed.* Slocan Park, BC: Silva Forest Foundation.

Harter, Susan. 1999. *The construction of the self: A developmental perspective.* New York: Guilford.

Harvey, L. D., and M. Reed. 1996. "Social science as the study of complex systems." In *Chaos theory in the social sciences*, ed. L. D. Kiel and E. Elliott, 295–323. Ann Arbor: University of Michigan Press.

Hearnshaw, E. J. S., R. Cullen, and K. F. D. Hughey. 2003. "Ecosystem health demystified: An ecological concept determined by economic means." A white paper, Lincoln University, Canterbury, New Zealand, http://een.anu.edu.au/eo5prpap/hearnshaw.doc.

Hendrix, Harville. 1967. "The ontological character of anxiety." *Journal of Religion and Health* 6, no. 1 (January): 46–65.

Hepworth, Dean, and Jo Ann Larsen. 1993. *Direct social work practice: Theory and skills,* 4th edition. Homewood, IL: Dorsey Press.

Holling, C. S. 1996. "Engineering resilience versus ecological resilience." In *Engineering within ecological constraints,* ed. P. Schulze, 31–44. Washington, DC: National Academy Press.

———. 1995. "Sustainability: The cross-scale dimension." in *Defining and measuring sustainability: The biogeophysical foundations,* ed. Mohan Munasinghe and Walter Shearer, 66–75. Washington, DC: World Bank.

———. 1973. "Resilience and stability of ecological systems." *Annual Review of Ecology and Systematics* 4:1–123.

Holling, C. S., D. W. Schindler, B. W. Walker, and J. Roughgarden. 1995. "Biodiversity in the functioning of ecosystems: An ecological synthesis." In *Biodiversity loss: Economic and ecological issues,* ed. C. Perrings, C. Mäler, K. G. Folke, C. S. Holling, and B. O. Jansson, 44–83. Cambridge: Cambridge University Press.

Johnson, Charles G. 1994. "Mountains: A plant ecologist's perspective on ecosystem processes and biological diversity." General technical report PNW-GTR-339 (September), United States Department of Agriculture (USDA), Forest Service, Pacific Northwest Research Station, Baker City, OR.

Kay, J. J., and H. A. Regier. 2000. "Uncertainty, complexity, and ecological integrity: Insights from an ecosystem approach." In *Implementing ecological integrity,* ed. P. Crabbe. Norwell, MA: Kluwer Academic.

Kessler, John T. 2003. "The healthy community movement: Seven counterintuitive next steps." *National Civic Review* 89 (3): 271–84.

Kierkegaard, S. 1944. *The concept of dread.* Princeton, NJ: Princeton University Press.

Kirchner, J. W. 1991. "The gaia hypothesis: Are they testable? Are they useful?" In *Scientists on gaia,* ed. S. Schneider, 38–46. Cambridge, MA: MIT Press.

Koestler, Arthur. 1990. *The ghost in the machine.* New York: Penguin.

Krauss, M. W., and F. Jacobs. 1990. "Family assessment: Purposes and techniques." In *Handbook of early childhood intervention,* ed. Samuel Meisels and Jack Shonkoff, 303–25, New York: Cambridge University Press.

Lang, Susan. 2000. "Home economics was a gateway for women into higher education, science careers." *Human Ecology* 28, no. 1 (September 22): 4.

Laughlin, Joan M. 2000. "Looking back to look forward: Reflections on women's issues (feminism?) in academe." *Clothing and Textiles Research Journal* 18 (3): 202–6.

Leider, Richard J. 2005. *The power of purpose: Creating meaning in your life and work.* San Francisco: Berrett-Koehler.

Lewis, J. 1979. *How's your family? A guide to identifying your family's strengths and weaknesses.* New York: Brunner/Mazel.

Light, S. S. 2001. "Adaptive ecosystem assessment and management: The path of last resort?" In *A guidebook for integrated ecological assessments,* ed. Mark E. Jensen and Patrick S. Bourgeron , 55–68. New York: Springer.

Lovelock, James. 2000. *Gaia: A new look at life on Earth.* New York: Oxford University Press.

———. 1988. *The ages of Gaia—a biography of our living Earth.* New York: Oxford University Press.

Madritch, Michael D., and Mark D. Hunter. 2002. "Phenotypic diversity influences ecosystem functioning in an oak sandhills community." *Ecology* 83 (8): 2084–90.

Mageau, Michael T., Robert Costanza, and Robert E. Ulanowicz. 1995. "The development and initial testing of a quantitative assessment of ecosystem health." *Ecosystem Health* 1:201–13.

Margulis, Lynn. 1998. *Symbiotic planet: A new look at evolution.* New York: Basic Books.

Margulis, Lynn, and J. Lovelock. 1976. "Is Mars a spaceship, too?" *Natural History* (June/July): 86–90.

Margulis, Lynn, and Dorion Sagan 1997. *Microcosmos: Four billion years of microbial evolution.* Berkeley, CA: University of California Press.

Marker, Sandra. 2003. "Unmet human needs." In *Beyond intractability,* eds. Guy Burgess and Heidi Burgess. Boulder: Conflict Research Consortium, University of Colorado, www.beyondintractability.org/essay/development—conflict—theory/.

Maslow, Abraham H. 1943. "A theory of human motivation." *Psychological Review* 50: 370–96.

Mayhew, Peter J. 2006. *Discovering evolutionary ecology: Bringing together ecology and evolution.* New York: Oxford University Press.

McCubbin, Hamilton I., and Joan M. Patterson. 1983. "The family stress model: The double abcx model of adjustment and adaption." In *Social stress and the family: Advances and developments in family stress theory and research,* ed. Hamilton L. McCubbin, Marvin Summan, and Joan M. Patterson, 7–37. Binghamton, NY: Haworth Press.

Minuchin, Salvador. 1998. *Family healing: Strategies for hope and understanding.* Cambridge, MA: Harvard University Press.

Mooney, H. A., and P. R. Ehrlich. 1997. "Ecosystem services: A fragmentary history." In *Nature's services: Societal dependence on batural ecosystems,* ed. G. Daily, 11–19. Washington, DC: Island Press.

Morin, Peter J. 1999. *Community ecology.* Hoboken, NJ: Wiley-Blackwell.

Muñoz-Erickson, Tischa A., Bernardo Aguilar-González, and Thomas D. Sisk. 2007. "Linking ecosystem health indicators and collaborative management: A systematic framework to evaluate ecological and social outcomes. *Ecology and Society* 12 (2): 6, www.ecologyandsociety.org/vol12/iss2/art6/.

Naeem, Shaid. 1998. "Species redundancy in ecosystem reliability." *Conservation Biology* 12:39–45.

Naeem, Shaid, et al. 1999. "Biodiversity and ecosystem functioning: Maintaining natural life support processes." *Issues in Ecology* 4:1–14.

Naeem, Shaid, L. J. Thompson, S. P. Lawler, J. H. Lawton, and R. M. Woodfin. 1994. "Declining biodiversity can affect the functioning of ecosystems." *Nature* 368:734–37.

Nisbit, E. G. 1991. *Living Earth: A short history of life and its home.* New York: Springer.

Norris, Tyler, and Linde Howell. 1999. *Healthy people in healthy communities: A dialogue guide.* Chicago: Coalition for Healthy Cities and Communities.

Nunes, Paulo A. L. D., Jeroen C. J. M. Van Den Bergh, and Peter Nijkamp. 2003. *The ecological economics of biodiversity: Methods and policy applications.* Northampton, MA: Edward Elgar.

Otto, H. 1962. "What is a strong family?" *Marriage and Family Living* 10:481–85.

Paul, Jeffrey. 1999. *Human flourishing.* Cambridge: Cambridge University Press.

Perrings, C. 1996. "Ecological resilience in the sustainability of economic development." In *Models of sustainable development,* eds. S. Faucheux, D. Pearce, and J. Proops, 231–52. Cheltenham, UK: Edward Elgar.

Perry, David A. 1995. *Forest ecosystems.* Baltimore: Johns Hopkins University Press.

Pimentel, David, Laura Westra, and Reed F. Noss, eds. 2000. *Ecological integrity: Integrating environment, conservation, and health.* Washington, DC: Island Press.

Pimm, S. 1991. *The balance of nature?* Chicago: University of Chicago Press.

———. 1984. "The complexity and stability of ecosystems." *Nature* 307:321–26.

Pittman, Thane S., and Kate R. Zeigler. 2006. "Basic human needs." In *Social psychology: A handbook of basic principles,* ed. Arie Kruglanski and Tory E. Higgins, 473–89. New York: Guilford.

Polanyi, M. 1968. "Life's irreducible structure." *Science* 160:1308–12.

Rapport, David J. 1998a. "Ecosystem health, ecological integrity, and sustainable development: Toward consilience." *Ecosystem Health* 4 (3): 145–46.

———. 1998b. "Defining ecosystem health." In *Ecosystem health,* ed. D. Rapport, R. Costanza, P. Epstein, C. Gaudet, and R. Levins, 18–33. Malden, MA: Blackwell.

Robbins, Michelle. 1996. "Thinking sustainably — sustainable ecosystems." *American Forester* 102, no. 2 (Spring): 33, http://findarticles.com/p/articles/mi—m1016/is—n2—v102/ai—18333986.

Rowe, Stan. 2001. "Transcending this poor Earth — á la Ken Wilber." *The Trumpeter* 17 (1): http://trumpeter.athabascau.ca/index.php/trumpet/article/view/138/161.

Salvesen, David, Lindell M. Marsh, and Douglas R. Porter. 1996. *Mitigation banking: Theory and practice.* Washington, DC: Island Press.

Scheff, Thomas J. 2007. "Universal human needs? After Maslow." In *Sociology at the frontiers of psychology,* ed. Gwynyth Overland, 19–39. Cambridge: Cambridge Scholars.

Scheffer, M., S. Carpenter, J. Foley, C. Folke, and B. Walker. 2001. "Catastrophic shifts in ecosystems." *Nature* 413:591–96.

Schindler, D. W. 1990. "Natural and anthropogenically imposed limitations to biotic richness in freshwaters." In *The Earth in transition: Patterns and processes of biotic impoverishment,* ed. G. Woodwell, 425–62. Cambridge: Cambridge University Press.

Schorr, Lisbeth. 1998. *Common purpose: Strengthening families and neighborhoods to rebuild America.* New York: Anchor.

Schulze, E.-D., and H. A. Mooney, eds. 1993. *Biodiversity and ecosystem function.* London: Springer-Verlag.

Schuster, P., and K. Sigmund. 1980. "Self-organisation of biological macromolecules and evolutionary stable strategies." In *Dynamics of synergetic systems,* ed. H. Haken, 23–39. London: Springer-Verlag.

Schweitzer, John. 1993. "Defining a healthy community." *Community News and Views* 6 (3): 1.

Stafford, L., and C. Bayer. 1993. *Interaction between parents and children.* Newbury Park: Sage Publications.

Stage, Sara, and Virginia B. Vincenti, eds. 1997. *Rethinking home economics: Women and the history of a profession.* Ithaca, NY: Cornell University Press.

Stinnett, N. 1979. "In search of strong families." In *Building family strengths: Blueprints for action,* ed. N. Stinnett, B. Chesser, and J. DeFrain, 23–30. Lincoln: University of Nebraska Press.

Taylor, Duncan M. 1994. *Off course: Restoring balance between Canadian so-*

ciety and the environment. Ottawa: International Development Research Center (IDRC).

Tillich, Paul. 1957. *Dynamics of faith*. New York: Harper and Row.

———. 1952. *The courage to be*. New Haven, CT: Yale University Press.

Tilman, Davie. 1999. "Biodiversity and ecosystem functioning: Maintaining natural life support processes." *Issues in Ecology* 4:1–11.

———. 1997. "Biodiversity and ecosystem functioning." In *Nature's services: Societal dependence on natural ecosystems*, ed. G. C. Daily, 93–112. Washington, DC: Island Press.

Tilman, Davie, P. B. Reich, J. Knops, D. Wedin, T. Mielke, and C. Lehman. 2001. "Diversity and productivity in a long-term grassland experiment." *Science* 294: 843–45.

Townsend, C. R., J. L. Harper, and M. Begon. 2000. *Essentials of ecology*. Boston: Blackwell Science.

United Nations (UN). 1998. *Universal declaration of human rights*. Fiftieth Anniversary of the Universal Declaration of Human Rights, www.un.org/Overview/rights.html.

Van Nes, Egbert H., and Marten Scheffer. 2007. "Slow recovery from perturbations as a generic indicator of a nearby catastrophic shift." *American Naturalist* 169:738–47.

Vernadsky, Vladimir. 1926. *The biosphere*. New York: Copernicus, Springer-Verlag.

Walker, B. H. 1992. "Biodiversity and ecological redundancy." *Conservation Biology* 6:18–23.

Walker, B. H., L. H. Gunderson, A. P. Kinzig, C. Folke, S. R. Carpenter, and L. Schultz. 2006. "A handful of heuristics and some propositions for understanding resilience in social-ecological systems." *Ecology and Society* 11 (13): www.ecologyandsociety.org/vol11/iss1/art13/.

Walker, B. H., C. S. Holling, S. R. Carpenter, and A. Kinzig. 2004. "Resilience, adaptability and transformability in social-ecological systems," *Ecology and Society* 9 (2): www.ecologyandsociety.org/vol9/iss2/art5/.

Walsh, F. 1982. *Normal family processes*. New York: Guilford.

Wessells, N. K., and J. L. Hopson. 1988. *Biology*. New York: Random House.

Wilber, Ken. 1996. *A brief history of everything*. Boston: Shambhala.

Wilkinson, David M. 2008. "Seven fundamental processes." Personal email communication to author, January 10.

———. 2006. *Fundamental processes in ecology: An Earth systems approach*. New York: Oxford University Press.

Williams, Garnett P. 1997. *Chaos theory tamed*. Washington, DC: Joseph Henry Press.

Williams, George Ronald. 1996. *The molecular biology of Gaia.* New York: Columbia University Press.

Wohl, Debra L., Satyam Arora, and Jessica R. Gladstone. 2004. "Functional redundancy supports biodiversity and ecosystem function in a closed and constant environment." *Ecology* 85, no. 6 (June): 1534–40.

Wolff, Tom. 2003. "The healthy communities movement: A time for transformation," *National Civic Review* 92 (2): 95–111.

Yaffe, Elaine 2005. *Mary Ingraham Bunting: Her two lives.* Savannah, GA: Frederic C. Beil.

Ziming, Liu. 1996. "Dissipative structure theory, synergetics, and their implications for the management of information systems." *Journal of the American Society for Information Science* 47 (2): 129–35.

Zimmerman, Michael E. 2004. "Humanity's relation to Gaia: Part of the whole, or member of the community?" *Trumpeter* 20 (1): 46–62.

Epilogue

Ansell, Chris, and Alison Gash. 2006. "Pragmatism and collaborative governance." Paper presented at the Public Administration Theory Network Conference, February 9, Evergreen State College, Olympia, WA.

Bowen, Murray. 1994. *Family therapy in clinical practice.* New York: Jason Aronson.

Coglianese, Cary. 2001. "Does consensus work? A pragmatic approach to public participation in the regulatory process." In *Renascent Pragmatism: Studies in Law and Social Science,* ed. Alfonso Morales. Hampshire, UK: Ashgate.

Gladding, Samuel. 2002. *Group work: A counseling specialty.* Upper Saddle River, NJ: Prentice Hall.

Jung, Carl. 1978. *Man and his symbols.* New York: Dell Press.

Lowen, Alexander. 1994. *Bioenergetics: The revolutionary therapy that uses the language of the body to heal the problems of the mind.* New York: Penguin.

Minuchin, Salvador. 1998. *Family healing: Strategies for hope and understanding.* New York: Free Press.

Muir, John. 1918. *Steep trails.* New York: Houghton Mifflin.

Ramirez, Efren. 1968. "The existential approach to the management of character disorders." *Review of Existential Psychology and Psychiatry* 8 (Winter): 43–53.

Roszak, Theodore. 2001. *The voice of the Earth: An exploration of ecopsychology.* Grand Rapids, MI: Phanes Press.

Satir, Virginia. 1988. *New peoplemaking.* Palo Alto, CA: Science and Behavior Books.

Shellenberger, Michael, and Ted Nordhaus. 2005. *The death of environmentalism: Global warming politics in a post-environmental world*. New York: Breakthrough Institute, Rockefeller Philanthropy Advisors.

Thomashow, Mitchell. 1996. *Ecological identity: Becoming a reflective environmentalist*. Boston: MIT Press.

Index

ahisma, 108, 109
Allah, 93–96
amala, 102
amana, 95
anomie, 20, 24; and Durkheim, 20
anthropic principle, 88–90
anthropocentrism, 3, 51, 54–55, 57–59, 205–206; necessary, 58–59, 60, 62–64, 157
anxiety: existential, 193; ontic and ontological, 193
aparigraha, 109
artha, 99
asteya, 109
aswada, 109
authority, and hierarchy, 140–142
autonomy: personal, 43–44; and social order, 43
autotroph, 128
Avatamsaka Sutra, 104

biodiversity, 127
bioenergetics, 199
biosphere: basic needs, 188–189; concept, 69; Earth as, 6, 129–130; Gaia, 183–185; planetary metabolism and ecology, 185–188; self-sustenance, 191–192
brahmacarya, 109
Brundtland Commission, 121
Buddhism, 55, 99–106, 111, 117; anthropocentric, 104; Christianity, 101; creation, 104–105; hierarchical levels of consciousness, 101–103; *Mahayanna*, 103; nature, 105; philosophy, 106; realms of existence, 104; rebirth, 103; religion, 106–107; transcendence, 106; variations in, 101

Canadian International Development Agency (CIDA), 154, 155
carbon sequestration, 129
Centers for Disease Control (CDC), 127
chi, 97–98
Christianity: ecological crisis, 85, 112–113; evangelical, 91; religious ecocosmology, 85–92; stewardship, 86, 91–92, 117–118
civic responsibility, 10, 205
collective unconsciousness, 199
Commission on the Ethics of Scientific Knowledge and Technology (COMEST), 71
commune, 36
communitarianism, 26–27, 37, 38, 44; approaches, 5, 43–48; environmental, 27; philosophy, 43; platform, 26
communities, 48, 137; "of communities," 10, 12, 35, 37, 41, 48, 157; confederation of, 48; functional, 165; healthy and sustainable, 165–166; rural, 41
community: capacity building, 165; commune, 36; communication and conflict resolution, 163–164; development, 39, 179–180; dimensions, 162–163; environmental sustainability, 164–66; healthy and sustainable, 120–122; importance of, 35–36; local, 25; perspective, 37–43, 163; provisioning and maintenance, 180–181

complementarity, 32–34
conflicts, mediating, 47
Confucian tradition, and eco-cosmology,
97–98, 111, 117
consensus, pragmatism and, 197–198
consumerism, 59
control, locus of, 19
cosmic ecology, 4, 6, 11–14, 67–118,
128–133; biosphere, 181; circumterrestrial space, 67–68; ecosystems, 11;
essential dimensions, 189–190; Gaia,
181–183; homemaking, 190; householding needs, 189; perspective, 69–
72; purposefulness, 74–76; unknown,
76–78
cosmology: complementary, 81–83; definition, 82; ecological, 81–118; and
reverential universe, 84–85; scientific,
81–84; spiritual, 81, 84–118
cosmos, lost *in* and *with*, 86
council of all beings, 56
courage to be, 195
culture of domination, 40

decomposers, 129
deep ecology, 6, 54–58, 112; definition,
55
dev, 108
dharma, 99–100

eco-fascism, 55
ecological action and intervention, 203–
205
ecologists: deep, 52; downstream and
upstream, 52–54
ecology, 126–128; affirmative, 5; cosmic,
4, 6, 11–14, 67–118, 128–132; deep,
6, 54–58, 112; definition, 4, 6, 14, 16,
38, 153; environmental, 5, 6, 10–11,
14, 50–54; environmentalism, 34;
family, 155–157; healthy, 119; "householder," 14; "householding," 13, 14;
human, 14–16, 35; humans, 3; nature,
3; needs, 147–149; nested, 4–6, 174,
188; obligations, 31; "of not," 5, 51–
54, 65; orientation, 20; personal, 5–9,
14–31, 119, 126, 149–152; place, 13;
reverential, 110, 111; self, 5, 19, 28–
30, 38; shallow, 55; social, 5, 9–10,
30, 38; shallow, 55; social, 5, 9–10,

15, 32–49, 64–66, 122–123, 152–
166, 180; solutions, 61–62; spiritual,
12, 14, 78; unknown, 5, 76–78, 81,
112, 116–118; utilitarian, 55
Economic Commission for Europe
(ECE), 126
economics, 41, 60; and behavior, 42–
43; free-market, 1; home, 152–155;
necessary, 58–59, 62; and sustainability, 2
ecosystem, 1; adaptive management,
142–143; autonomy, 174; biodiversity,
127–128; composition and distribution, 167; constancy and stability,
171–173, 176; disturbance, 168; diversity, 170–171; frequency and amplitude, 173; functional redundancy, 170;
health, 119, 180; invigoration, 168;
natural, 48; organization, 167–168;
persistence, 171; perturbation and
rebound, 169, 171; resilience, 127,
168–170, 176; species and structural
stability, 172–174, 176; sustainability,
173–176; temporal elasticity, 168;
vigor, 127, 166–167, 174, 176
education: environmental, 8; system, 22
energy: chaos and disorder, 176; entropy,
175–176; flows and transfers, 128,
175–176; negative entropy, 175;
resources, 176; second law of thermodynamics, 175
entropy, 175–176, 194; negative, 175
environment: built, 60; and control, 19;
and ecology, 5, 10–11, 24; and education, 8, 198–200; and health, 119;
healthy and sustainable, 126–128; and
the past, 18
environmental ecology, 5, 6, 10–11, 14,
50–54; affirmative, 63–64; basic
needs, 166; definition, 51, 54; essential
dimensions, 177–178; healthy and sustainable, 126–128; householding
needs, 176–177; noise level, 173
environmentalism, 200–203; death of,
201
environmental movement, 112
ethic: ecological, 8, 116; land, 26–27
evangelism, 116
exosystems, 7

family, 47; adaptation, 123; clinicians, x; community, 153, 162–166; dysfunctional, 122; ecology, 155–157; health, 119; healthy and sustainable, 122–124; home economics, 154–162; householding, 160; households, 179; identity and interpersonal interaction, 159; leadership, 123, 158–159; needs and functions, 157–163; procreation and parenting, 158–159; resource management and procurement, 162; spiritual nurturing, 160; support, 158–159; system boundary maintenance and integration, 160; system communication, 161–162; systems, x; therapy, 122
fasad, 94
fitra, 94

Gaia, 128; biosphere, 181–183; holon, 183–185; living organism, 182; spaceship earth, 183, 192; super-organism, 183–185
Garden of Eden, 85
gati, 108
golden rule, new, 46
good enough, 61–62

health: and community, 165–166, 120–122; ecological, 119, 180; environments, 127; family, 122–124; lifestyles, 124–126; spiritual, 128–133; sustainable, 120; systems theory, 9
health services, 154–155
hierarchy, 33; authority, 141; Chinese boxes, 141; definition, 140; formal, 141; nature (natural), 140, 142–143; needs, 3, 4, 144–145, 147–149; nested, 6–7, 141–143; nested systems, 138–139; relationships, 36; social, 139; subordination, 140–141; systems, 6, 7, 140–144
Hinduism: Buddhism and Jainism, 100, 111; deep ecology, 99; eco-cosmology, 98–100; environmentalism, 113–114, 117
Hindutva, 115
holon, 183–185
home, 149–152

homemaking, 178–179
households and householding, xi, 13, 14, 38, 137, 141, 149–150, 151, 177, 179; human, 190–191
human nature, 44–46, 59–61, 63; communitarian perspective, 46–48; dour, 45–46, 56–57, 197; sanguine, 44–45, 197
humans: development, 15; ecology, 14–16; flourishing, 155; fulfillment, 31; needs, 24, 154; satisfaction, 31; threat, 55; virtue, 47; wants, 24
hypercycles, 129

identity, 198
indemnity: place, 17–19; self, 17
individualism and individuality, 28, 40, 44
Institute of Medicine (IOM), 127
interdependency, 27–28
internalization, 47
introspection, 199
Islam: eco-cosmology, 93–96; and scientific discovery, 96; stewardship, 93–96, 112

Jainism: Buddhism and, 108, 111; cosmology, 108–109
jivas, 108
Judaism, 117

kalipha, 94
kama, 99
karma, 99, 102, 109
khilifa, 95

Land Ethic, 26–27; definition, 27
Leopold, Aldo, 143–144
lifestyles: behaviors, 125; economic resources, 125; healthy and sustainable, 124–126, unhealthy, 124
love, 147

macrosystems, 7
management, ecological orientation, 16
manas, 102
mesosystems, 7
microsystems, 7
mixed scanning, x
mizan, 95

moksah, 108, 109
morality, 47–48; infrastructure, 47;
    voice, 48
moska, 99
mountain, thinking like a, 143

naraki, 108
National Center for Environmental
    Health (NCEH), 127
Native American religion, 55
natural step, the (TNS), 52–53
needs: basic, 25, 144–195; hierarchy, 3,
    4, 144; higher and lower order, 3
neighborhoods, 180
neo-paganism, 115
nesting: domains, 8; ecology, 4–6, 174,
    188; hierarchy, 6–7, 141–143; infra-
    structure, 48; interrelationships, 156;
    systems, 6
nirvana, 101
nivritti, 99
nomads, 25
nostalgia, 42
nuclear power, 72
nuclear waste, 72–73

other, 29–30; spirituality and, 30–31
outer space: debris, 67–68; ethics, 71–
    72; nuclear waste, 72–73

panarchy, 143
paradise, 86
parasites, 129
personal level: basic capacities, 195; ecol-
    ogy, 5–9, 14–31; health, 119; health
    and sustainability, 126; householding,
    149–152; needs, 147–149; orienta-
    tion, 20
photosynthesis, 129
place, 150; bound, 24; identity, 18, 19;
    portable sense of, 21; rightful, 28;
    sense of, 17, 20, 21–24, 26–27, 42;
    specific, 27
populations, human, 120
pragmatism, 62–64, 133–138; anthropo-
    centrism, 135–136; assumptions, 137;
    collaboration, 196–198; definition,
    134; ecological worldview, 137–138;
    nested ecology, 135–138; traits, 196

pravritti, 99
precautionary principle, 60–61
preservation, ecological, 76
principles, first order, 52–54, 65
productivity, ecological, 127
psychology: approach, 203–204;
    developmental, 15; formation, 19;
    social, 17
psychotherapy, 198–203
purusharthas, 99

Quran, 93–96

rationality, limitations of, 76–79
religion, 116–118, 194; discipline, 111;
    intolerance, 79–80; traditions, 111–
    115. *See also* Buddhism; Christianity;
    Confucian tradition; Hinduism; Islam;
    Jainism
resilience, 127–128
resources, non-renewable, 180
responsibility, 179, 192; civic, 10, 205;
    ecological, 146–147; housekeeping,
    163
reverence: ecology, 110, 111; universe,
    84–85, 110
rights, basic, 154–155; ecological, 31

samsara, 99
sarvadharma samantava, 109
sarvatra bhaya varjana, 109
satisficing, 42–43, 51, 61–62, 197; defi-
    nition, 42
satya, 109
self: actualization, 148, 149; awareness,
    29; care, 111; concept, 17; conscious-
    ness, 39; ecology, 5, 19, 37, 38, 42;
    fulfillment, 15; identity, 17–18;
    involvement, 29; moral, 47; and oth-
    ers, 28–30, 109, 111; preservation, 76;
    projection, 151; realization, 39; reli-
    ance, 41; socially competent, 36;
    socially emergent, 29; sufficient, 41
shariashram, 109
social ecology, 5, 9–10, 32–49, 54, 64–
    66, 122–123, 152–166, 180–195;
    basic needs, 152–155; definition, 32–
    36, 38; family needs and functions,
    157–163

social formations, 47

social intercourse, 36

social system: characteristics, 156; theory, 7

society, temporary, 21–23, 25–26

solar energy, 70

space: circumterrestrial, 68–69; waste, 72–74

spaceship earth, 192

sparsha bhvana, 109

spirituality, 30–31, 194; common good, 132; faith, 131; healthy and sustainable, 128–133; metaphor, 133; otherness, 132; perspective, 30–31; religion or theology, 77–80; transcendence, 132

standard theory, 83, 84, 111

stewardship, 27–28, 35, 78

stress, resistance to, 127

sustainability, 9, 11, 24, 25, 28, 31, 34, 42–44, 53–54, 56, 117; definition, 121–122; and health, 119–120

swadeshi, 109, 111

systems: open and closed, 174–176; terrestrial, 70–71

systems theory: definition, 7; general, 6; health, 9; sustainability, 9

tiryanc, 108

tradeoffs, 129

twahid, 94

United Nations, 154, 155

universes, 12

unknown, ecology of, 76–78, 81, 112

utilitarianism, 51; ecology, 110; necessary, 58–59, 60

vaidika dharma, 99

values, 111

veda, 99–100

veil of karma, 109

World Health Organization (WHO), 120–121, 127